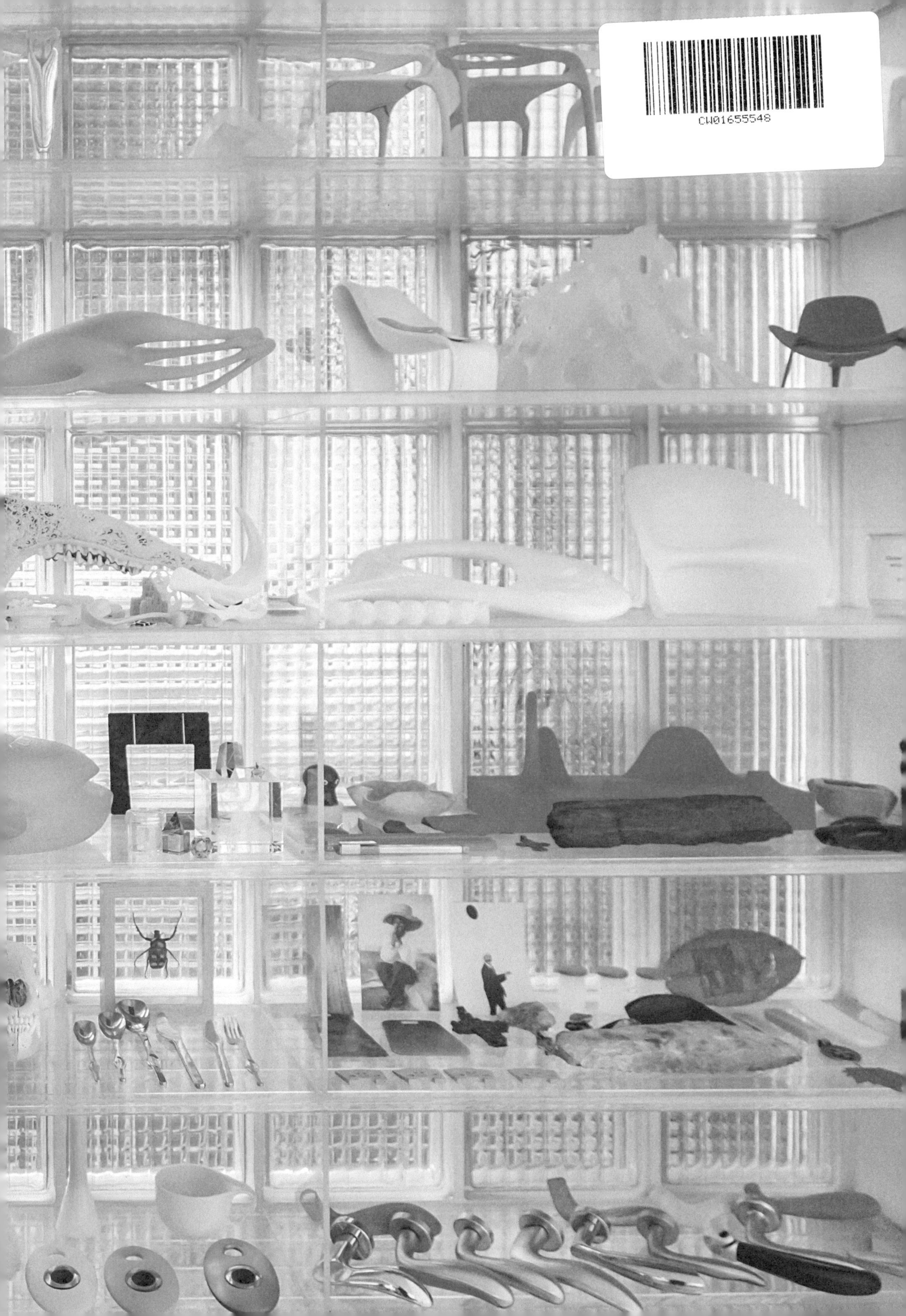

Ross Lovegrove

Convergence

Sam, Kim

with total respect.

Ross. 25. July 2018
studio

Edited by
Marie-Ange Brayer

SIEVEKING
VERLAG

Centre national d'art et de culture Georges Pompidou

The Centre National d'Art et de Culture Georges Pompidou is a national public establishment under the supervision of the Ministry of Culture (Law 75-1, January 3, 1975).

Marie-Ange Brayer
Curator of the exhibition, Head of Industrial and Prospective Design

The exhibition *Ross Lovegrove: Convergence* was made possible by the generous support of

SWAROVSKI
Major exhibition partner

ENEDIS
L'ELECTRICITE EN RESEAU
Major sponsor of the Centre Pompidou

Schweizerische Eidgenossenschaft
Confédération suisse
Confederazione Svizzera
Confederaziun svizra
Ambassade de Suisse en France

and of

LASVIT BARRISOL Artemide KEF
RENAULT MOROSO yamagiwa

Contents

Acknowledgments

This exhibition and publication would not have been possible without the exceptional commitment of Ross Lovegrove. Our special thanks go to him, and to his main partners, for their invaluable help during this project, Charline Ronzon-Jaricot and Stefan Bassing from the Ross Lovegrove Studio, as well as to his team: J. J. Lee, Pierre Bacaud, and Raquel Fernandez.
We would particularly like to thank Frédéric Migayrou, deputy director responsible for Industrial Creation, who instigated this exhibition project. We also warmly thank Ilaria Colombo for her support.
We are also deeply grateful to all the sponsors and partners who have supported the exhibition: Nadja Swarovski, to whom we express our sincere appreciation; Swarovski and its directors: Robert Buchbauer, Mathias Margreiter, Markus Langes-Swarovski, Daniel Cohen; as well as Julie Martinez, Marjan Tharin, and Lila Thibault for their involvement in the exhibition;
Barrisol Normalu and its chief executive, Jean-Marc Scherrer, as well as Philippe Mahieu, Edwin Kouijzer, Paulo de Melo, and José Soriano for their invaluable help;
KEF and Grace Lo, as well as its founder, Raymond Cooke, along with Jörgen Halldin, Ray Fok, David Leung, Linda Yau, and Marie Yuen for their vital assistance;
Artemide and its chief executive, Ernesto Gismondi, its deputy chairman, Carlotta de Bevilacqua, and Alain Taillandier, managing director of Artemide France, as well as Diego Martinelli;
Lasvit and its chief executive, Leon Jakimic, together with Michal Supita and Štěpánka Barešová, Mirka Horáková, Hana Klímová, Pavel Krajčík, Michaela Losenická, and Ales Styblo, for their help with this project;
Renault and its chief executive officer, Carlos Ghosn, as well as Laurens van den Acker, Anthony Lo, and Olivier Rouxel, for their support and availability;
Moroso, its chief executive officer, Roberto Moroso, artistic director Patrizia Moroso, in addition to Alice Caudera, Marco Capellini, Concetta Giannangeli, Davide Michelizza, and Alessandra Monteleone; we would also like to thank Stratasys for contributing expert knowledge during the creation of the Spine Chair manufactured by Moroso;
Materialise and its directors, Wilfred Vancraen and Bart Van der Schueren, as well as Andrew Jones for his assistance with the exhibition.
We particularly want to thank Mitsuhiko Endo and the firm Yamagiwa.
We would like to express our gratitude to the partnering brands and manufacturers of the Ross Lovegrove works for their cooperation and their loans: Kiko, Luceplan, Vondom, and United Nude.
We also thank Alisa Andrasek, David Chipperfield, Isaie Bloch, the White Russian Studio, and Alain Prochiantz.
We are deeply indebted to the Muséum national d'histoire naturelle in Paris for their scientific collaboration, especially to André Nel from the entomology laboratory, as well as to Claude-Anne Gauthier, director of the Galleries Department at the museum, Cécile Colin, manager of the Comparative Anatomy and Paleontology galleries, Christine Argot, in charge of mammal fossil conservation, and Camille Wolff-Dalem, for allowing us shoot the film for this exhibition in the Paleontology gallery.
We warmly thank all the people at the Centre Pompidou who, in one way or another, lent their invaluable help to this project, particularly Marie-Annick Hamon, in charge of production, for her exceptional contribution, and Geneviève Munier, subeditor.
Finally, our warmest thanks go to the authors of this catalogue, Ross Lovegrove, Marie-Ange Brayer, Frédéric Migayrou, and Olivier Zeitoun, as well as to Mathilde Vallée and Marie Siguier, for their research.

Acknowledgments

Ross Lovegrove

I would first of all like to thank:
Miska Miller-Lovegrove, architect, who, after having championed the project over several years, made the present exhibition at the Centre Pompidou possible and who has indeed unselfishly supported my work as a designer in all senses of the term, along with establishing our studio and supporting me with a vision of design tied to art and culture;
Roman Lovegrove, whose integrity of conceptual thought in the field of contemporary architecture is a source of inspiration to me; he inspired the second generation at Villa Matoso and has lent the model for the present exhibition.
I am deeply grateful in particular to my devoted assistants, who helped me to identify the key works shown in the exhibition:
Stefan Bassing, for having coordinated the exhibition with so much intelligence and rationality, which allowed the project to be completed, and for having thought of the very beautiful *Barrisol Cocoon* for *Instinctive Override*;
Charline Ronzon-Jaricot, for having coordinated the exhibition with incredible dedication, watching over the smallest details of its realization, becoming my "hard drive" that brought together all the knowledge necessary to make the presentation a comprehensive exhibition;
Christoph Herman, who, with his extraordinary attitude, was able to elevate our work to a higher level of calculative design, passing from architecture to industrial design, in projects that were all manufactured without aesthetic or conceptual compromise: Lasvit LiquidKristal pavilion, Nebula, Foliates, Chlorophilia, Florensis, Corolised, Diatom, Renault Twin'Z and the Kiko collections;
Pierre Jusselme, for his exceptional contribution to my work in all respects and, more particularly, during the creation of the Swarovski Crystal Aerospace, Solar Tree, and GT110 projects, along with many others;
Arturo Tedeschi, the world leader in the use of Grasshopper software, who, with Ilaria Colombo, got involved, with enormous creative generosity, in the making of the Ilabo shoes for United Nude, which has since become a point of reference in the realm of industrial 3-D printing;
Tim Williamson, for having collaborated with me on all the projects that define me, Tŷ Nant, DNA, Muon, Bioform, and Gingko at Novartis, and on so many others during this exceptional decade;
Gianluca Berruti, for his inspiring cooperation on *Instinctive Override* and *Monolith* shown in the exhibition, the most rewarding projects of my career;
Matt Longbottom, for having translated my sculptural projects and data in order to create the gentle lines of Moot, Biolove, Twin'Z, and Factoria;
Anka and Sean Simone, for the communication and production of the film;
John Ross, for the photography;
as well as the team members who helped me on earlier important projects: Julia Koerner, Anna Sturmberger, Gernot Oberfell, Jan Wertel, Sebastian Liedke, Christoph Behling, Tilla Goldberg, Christoph Moinat, Joe Heapy, Andrea De Benedetto, Mathius Barr, and Hiromichi Konno.

Foreword

Serge Lasvignes
Chairman, Director & CEO, Centre Pompidou

Bernard Blistène
Director, Musée national d'art moderne –
Centre de création industrielle

To establish a new dialogue between nature and technology, making the arts and sciences converge, the British designer Ross Lovegrove has adopted, with a profound originality, a holistic vision of design. As early as the 1990s, his visionary work practice took into account digital evolutions. At the same time, it rejected the productivity logic of mass industry by substituting an economy of materials and forms. This first major solo exhibition that the Centre Pompidou is devoting to Lovegrove highlights the position of design in the postindustrial era, a juncture at which a significant shift from the mechanical to the biological is taking place, a transformation symptomatic of our time that emphasizes "digital forms." As such, this exhibition naturally positions itself at the heart of the new annual multidisciplinary event "Mutations-Créations," which the Centre Pompidou is symbolically launching in the year of its fortieth anniversary, with the wish to reactivate the spirit demonstrated at the outset by the Centre de création industrielle.

From the designer's most emblematic objects to works in progress created specifically for the exhibition, visitors will discover Lovegrove's unique approach through more than one hundred design objects, prototypes, drawings, and installations, displayed in an exhibition space specially created by the designer.

Kinetic "technological sculptures," mutant vehicles propelled by the movement of a drop of water, parametric glass architecture, cosmic lamps, and other "naturalized" everyday objects (bamboo bicycles and aquatic-shaped water bottles) are presented in dialogue with natural-history specimens and other artifacts that inspire Ross Lovegrove's work and contribute to making an oeuvre directly linked with the Zeitgeist.

We warmly thank Ross Lovegrove for his generosity, his humanity, and his complete commitment to the creation of the exhibition. We extend our thanks to his studio and in particular to Charline Ronzon-Jaricot and Stefan Bassing. We would also like to acknowledge the remarkable involvement of Marie-Ange Brayer, curator of the exhibition and head of Industrial and Prospective Design, as well as that of Olivier Zeitoun, deputy curator, who successfully realized this ambitious project.

Finally, the exhibition would never have been possible without the exceptional support of Nadja Swarovski (Swarovski), as well as the invaluable cooperation of Philippe Monloubou (Enedis), Ambassador Bernardino Regazzoni (Swiss Embassy in France), Leon Jakimic (Lasvit), Jean-Marc Scherrer (Barrisol), Grace Lo (KEF), Laurens van den Acker and Anthony Lo (Renault), Patrizia Moroso (Moroso), Carlotta de Bevilacqua (Artemide), Mr. Endo (Yamagiwa). We are most grateful to them all.

Mutations / Créations

Frank Madlener
Director, Ircam

Frédéric Migayrou
Deputy Director, Musée national d'art moderne – Centre de création industrielle

Marie-Ange Brayer
Curator, Head of Industrial and Prospective Design, Musée national d'art moderne – Centre de création industrielle

The Centre Pompidou introduces a new event, titled "Mutations/Créations," which focuses on the future potential and interactions between digital technology and creation; a domain shared by art, innovation, and science. Cross-functional and interdisciplinary, mixing research, the arts, and engineering, this annual event brings together, for its premiere occasion, music, design, and architecture. It offers two exhibitions—*Printing the World* and *Ross Lovegrove: Convergence*—along with the art and innovation forum "Vertigo," as well as study days and meetings.
Each year, thematic and monographic exhibitions will be supplemented by a program of meetings and workshops, which will turn the Centre Pompidou into an "incubator," a venue for demonstrations of prototypes, live artistic experiments, and discussions with creators. This platform will also serve as a critical observatory and a tool for analyzing the impact of creativity on society.
How does the creative community use digital technology to open up new industrial perspectives? How do they examine the social, economic, and political phenomena and the ethical limits of these industrial developments? What are the formal evolutions in the areas of music, art, design, and architecture in light of the advances in technoscience?
We can safely bet that "Mutations/Créations" will offer many diverse and illuminating answers to these major questions.

Designing Evolution

Ross Lovegrove

The imagination is a force of nature and a constant catalyst for the stimulation of progress and biodiversity on our rare and beautiful spaceship Earth. . . . It is something truly remarkable, an optimistic and inspiring force that can fuse together logic and beauty in the evolution of intelligent life-forms.

The systems that have existed for eons within the context of primary evolution developed harmonically through the effects of natural selection, chance, mutation, atmospheric conditions, and a slow burn in time. Today these phenomena are being replaced by artificial selection in a revolutionary new phase of tertiary evolution that is accelerating innovation and creation toward endless possibilities.

The recompositional, speculative power of the human mind, now augmented by artificial intelligence, is turning thought into reality.

As such, the question emerges of whether design itself will remain a conscious, human investigation, or become subject to systematic algorithmic premonition in search of optimized solutions. I recently posed this question to Richard Dawkins, the preeminent evolutionary biologist of our time: "If the biological mind can conceive and construct machines of hard physicality, does this mean that to some degree they are biological or embody biological heritage?"

Dawkins replied that this was my subject, not his, which is a response that tends to suggest that evolutionary biology is separated from conscious, hard-copy creation—a notion with which I fundamentally disagree. Design certainly has a much broader bandwidth of meaning, and in my view, it is a subject that represents an agent of change, as if there was indeed a gene for creativity that some individuals have and others do not.

There is a convergence between evolutionary design on the one hand, and creation theory on the other, because when theorists and philosophers hit a brick wall they tend to accept that life is so complex it must have been created by another entity rather than by chance and glitch.

Perhaps nature itself should be regarded as a force, just as gravity and magnetism are.

However clearly we break down and understand the building blocks of life at a subatomic level, advancing software allows us to design in an unrestricted biomorphic way, blending and coalescing materials and forms almost anatomically, in every sphere of the physical

world around us; then we are surely evolving toward an organic future that includes the transformation of everything, including ourselves.
In support of this idea is the exponential growth of knowledge, defined by Moore's Law, that is moving us rapidly from a mechanical system to a biological one, as the shift from the analogue to the digital is proving both revelatory and succinct, leading to the exploration of the third and fourth dimensions of everything around us, from medicine, biotechnology, and science to space, architecture, materials innovation, design, and art.
This computational realm is transformational for life on Earth. It harnesses the benefits of systems efficiency, augmented, in the case of cars for example, by satellite navigation, embedded artificial intelligence, and proximity sensors. It is leading us toward a new form of collective human experience that is possibly sustainable if our deeper environmental consciousness through the World Wide Web can help us to be more acute and Earth-centric in our overall collective awareness of both micro and macro conditions that effect our collective growth and survival.
In terms of generating ideas and concepts, we should be searching for and elevating individuals with a polymathic mind, that is, those rare individuals with the ability to link seemingly unrelated aspects of life and guide the possibilities by way of maintaining an instinctive awareness of all things considered simultaneously.
This story began in the 1850s, a crucial decade marking the outset of the convergence during which the interconnections and overlaps between the various disciplines first started to reveal themselves in two fundamental areas and thus added a whole new dimension beyond science, one that had not been fully grasped until then. The idea of energy conservation was first established, which brought together contemporary discoveries in the sciences of thermodynamics, optics, electricity, magnetism, and biochemistry.
Almost simultaneously, Darwin's theory of evolution by natural selection brought together the new sciences of deep space astronomy, deep time geology, paleontology, anthropology, geography, and biology. These years saw the greatest intellectual breakthrough of all time, the realization that one science supports another. They marked the beginning of a form of understanding like no other and, from an intellectual perspective, a new era.
I sense that we have entered a new era of convergence and that multiple possibilities for advancing our civilization now lie in the associations and affiliations between the contemporary equivalents of Darwin's references, extending the scope of exploration exponentially. Convergence is a remerging order that is so strong and coherent that it is infiltrating other systems of knowledge, traditionally different from or even

opposed to science, and is starting to explain and advance them. The methods and the infrastructure of science are invaluable; indeed, they are unrivalled aspects of democracy and will shape society in all its manifestations even more in the future than they have in the past.
All the same, we have designated human beings as the principle species to be given special veneration; we have set this designation at the very core of our reasoning. But in doing so, we are becoming blind to the effects that exponential growth at odds with our biosphere can have on our ecosystem, and to the subsequent depletion of other species comprising our collective support system.
Anyone can make anything with any resources they feel free to find or acquire, by whatever means, without any real responsibility toward humanity as a whole.
We are evolving a very different set of conditions in our newly defined Anthropocene, an era in which industry and the making of things have to establish new laws in respect of cohabitation and a sustainable lifecycle of resources.
Today, as the forces of good and evil battle it out like avatars, we see a positive and optimistic hope in the quest for an alignment with nature, as articulated by new sciences such as biomimetics and the new additive manufacturing techniques, which are advancing rapidly in the creation of multidimensional structures across materials, scales, and disciplines.
We need to shift design from what "looks" modern to what "is" modern—without being prescriptive in the process—and understand what being progressive really means.
Ironically, the new tools and processes being developed for the new age of manufacturing will create an incredible blossoming of diversity in the organic, biomorphic, biological sense, in sync with nature, so that the physicality of everything around us is an intelligent evolutionary promotion of our world centered on the earth as well as on humans.
Divisions between natural and artificial objects will blur as the principles of nature are transmuted into everything from the chlorophyllic, high-efficiency, nano solar skins of cars and buildings to the coatings of bridges and motorways. The generation of energy itself will undergo a major revolutionary transformation such that energy will be produced directly on a microatomic level (fusion), creating an off-grid level of mobility and independence globally, without atmospheric pollution and its threat to life as we know it.
Within the next fifteen years we will already begin to look back and wonder at the crudeness, weight, brutality, excess, and imposition of the mechanical designs we are creating today, as everything is

set for reengineering and redesigning in the biological sense, with a fat-free economic mandate for the convergence of intelligent technologies.

I see around me signs of the emergence of a new shape-shifting force that is already challenging the mainstream: labs creating contact lenses that will provide us with shared vision and a direct link to the Internet; advances in genomic medicine, prosthetic augmentation, mapping of the brain; or major twenty-first-century corporations investing in fitness and a new awareness of nutritional health and well-being, while thousands of new start-ups are creating fabulous new products made possible by the newfound freedom of mating hardware and software. Here we could cite as an example Emotiv Insight's incredible headset that reads brainwaves and can turn a visualized image into a 3-D printed object. Incredible . . .

But new ideas need new forms and new languages to amplify the paradigm shift and keep the human mind vital and inspired by change; to help us remain curious but at the same time not weighted down by the gravity of industrial physicality we are experiencing now, as in cars or computers. Hence, the role of the designer also needs shift to remain relevant and aligned to the potential of the age in which we live. This will offer designers the potential to create with great succinctness and purpose, promoting values that are collectively vital to us all. In doing so, they will also influence a new, emergent generation of creative minds endowed with a passionate understanding of nature and its intelligence.

Design in the twenty-first century leans forward in anticipation of change; it is a form of fusion in itself, whereby innovation, invention, science, technology, emotion, and humanism meet with reverence for our beautiful planet. We now live in a scenario that is very different from that of the twentieth century, which saw a handful of designers and architects embrace technological innovation to propose a "brave new world" that resembled a dream of living in an advanced civilization with highly efficient systems, lightweight structures, and streamlined transportation networks.

Speculating on the future is a beautiful and enriching experience as we try to anticipate and then bring to life the dreams and visions that fuel our economies and, in our own way, modernize, harmonize, and bring greater value to life.

The key to the future, however, is the retention of our instinct and our wisdom, both of which might fail in the face of the onslaught of artificial intelligence and virtual reality. The saddest aspect of society right now is that science gathers knowledge faster than it gathers wisdom. Unlike information, wisdom cannot be dispensed via blogs and Internet chatter. Since we are drowning in an ocean of information, the most precious commodity in modern society is indeed wisdom. Without wisdom and insight, we are left to drift aimlessly, lacking an objective purpose and harboring a hollow feeling, once the novelty of unlimited information wears off.

Design is a seemingly endless release of ideas and concepts; it is an agent of change and diversity, but we must begin to understand the layers of design and their hierarchy in order to promote a new age of intelligent, Earth-centric enlightenment. The way in which we grant absolute value to a harmonious coexistence between biosphere and pure, fundamental human emotions will therefore allow us to create and ultimately preserve that coexistence.

London, December 30, 2016

The Nature of Synthetic Design

Frédéric Migayrou

My theory of the formation of the scientific doctrine of matter is that first philosophy illegitimately transformed the bare entity, which is simply an abstraction necessary for the method of thought, into the metaphysical substratum of these factors in nature which in various senses are assigned to entities as their attributes.

Alfred North Whitehead[1]

There is an evident phenomenality in Ross Lovegrove's creations, for they immediately impose the idea of a fluidity, of a movement in which objects progressively reveal themselves in the diversity of the approaches and successive perceptions of contours. Here, form yields itself in sequences, renews itself with changes in lighting effects, multiplies into something cinematic, like the suspended luminaries in the Barrisol Pavilion (2014) that lent a plasmatic dimension to the space. Beyond its function and usability, the object is always a generator that immediately interacts with its environment, an entity that seems to absent itself the better to make its presence felt, following the rigorous definition of a few contours, in an *épure*, a schematic drawing whose lines are prolonged and blend into the surroundings. The exhibition *Endless* at the Cardi Black Box gallery in Milan (2008) accentuated, through a multiplication of reflections, this continuity between matter and space, the forms radiating with luminescence. This work of reduction, of elision, gives the objects a powerful physical identity, one of forms so evident they evoke a universe of archetypes, of ancestral morphologies, of the material world and living organisms, of representative universal forms that touch us, that strike the collective unconscious,

Spongy bone seen under the microscope

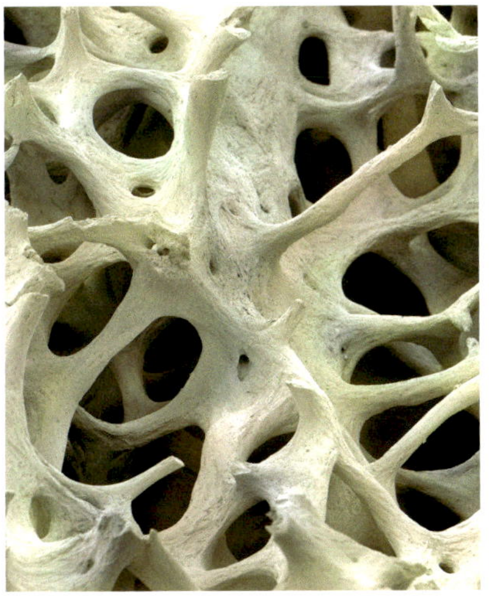

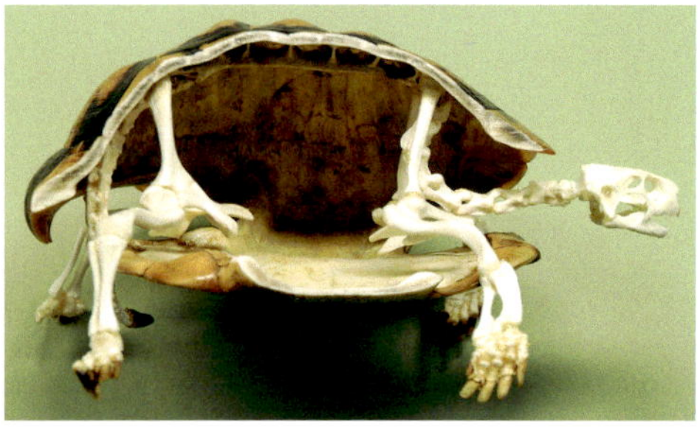

Tortoise
Cross section of the carapace and the skeleton

our reptilian memory, the *facultas praeformandi*, or possibility of representation, that Carl Gustav Jung imagined structured our primal consciousness, drawing from the sources of the psyche, as much human as animal.

Through this search for universal forms, the man who has described himself as a "sculptor of technology" delineates his area of practice: namely, a form of industrial design in agreement with a new logic of production and industrialization that aspires to go beyond the notion of product and to escape a critical economy of the object, a type of postmodernism, an appropriation-based design made of images and assemblages that, by preserving the distance of a subjectivity, endorsed an ontological status of the industrial object, and therefore an ultimately very conventional idea of function and use. Against the backdrop of radical design, or the cynicism of a certain type of critical design based on narratives (or design fiction), Lovegrove owns his image as a positivist attentive to the hypotheses presented by futurology. Working with new materials and digital technologies, he neutralizes the univocal logic of progress by bringing any future back to transhistorical models rooted in the deepest depths of human cultures. Lovegrove eschews a culture of design that assumes, in one way or another, that the act of conception remains the preserve of an intentional dynamic on the part of the designer, architect, or engineer as the creator of forms, environments, and scenarios. He privileges a stance that has more to do with ergonomics, with a permanent adaptation of objects, environments, and physical spaces to human activities. This is no longer simply a matter of arranging artifacts, instruments, the machines of architecture to enhance productivity. Rather it entails using, in a world entirely governed by industrialized procedures, methods that, to redefine the conditions of adaptability of humans to artificial universes, have opened onto other disciplines: engineering science, management, sociology, but also psychology or physiology. Design can thus be attached to sets of interfaces, be they material, technological, or cognitive.

With Lovegrove, industrial aesthetics reconnects with this initial sense of ergonomics, the one that Ernst Haeckel gave it in introducing the word *ergology,* and then ecology to underscore the capacity of living organisms to adapt to their environment. His body of work reconciles design with the primitive intelligence of *oikos* (in Ancient Greek, the house, the family property).

Suction-cup foot of gold dust day gecko from Madagascar

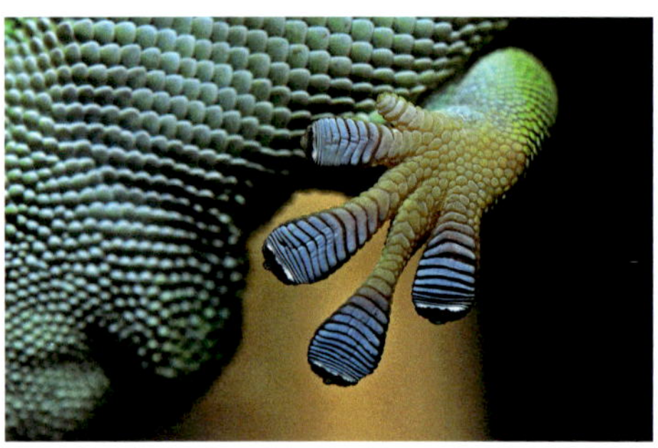

To lend shape consists therefore in revealing, in optimizing a form replete with multiple tensions, external determinations, and constraints, where the object is the outcome of a play of interactions between the physical characteristics of the materials, and between the forces and movements emerging from the functions, conditions of use, and overall economy of the project. The reason Lovegrove evokes liquid, fluid universes so often is because he finds therein the logic that corresponds best to his approach to creation, one that invites him to rely entirely on morphogenetic imperatives. One might also think of the field of geomorphology, where physical forces and tensions make contours and masses surface, suggesting the unity of a form generated by the parameters of a context and a situation. This is the case of the Lasvit LiquidKristal pavilion (2012) and its undulating glass surfaces that play with transparency so as to transmute surrounding bodies and objects into motion. On an architectural scale, the glass screen becomes the substratum of a phenomenal dynamic that gives rise to endless compositions. Much earlier in his career, after more than a year of research, Lovegrove managed to capture in a few lines the movement of a trickle of water to create a bottle for the company Tŷ Nant. From the diagrammatic section indicating the complex variability of the contours to the composition of the surfaces, the water bottle looks like an envelope, an invisible skin beneath which the water displays its dynamic and optical properties by materializing the form of the container. All of Lovegrove's creations respond to this economy of a delineation of a serpentine figure where curves and counter-curves are entwined, and the three-dimensional volume resulting from this reduction to a surface, to an infrathin, create, "according to a balance of intrinsic and extrinsic forces, the form's natural scale." From this play on restriction, this search for the most tenuous, most determined line, for a line tending toward the *épure*—*épure* understood in its original sense, that of a disquisition of the elementary purity of a drawing for the purposes of representing a volume—Lovegrove has established the foundations of a methodology in which reduction finds a critical function, opens the way to a questioning of the origin of forms, of access to morphological constants constituting the structures of the material world but also of the organic world. To use a term borrowed from phenomenology, it is a matter, by leaning on an *epochè*, of bracketing

Knoll's ad for Eero Saarinen's Tulip Arm Chair
1969

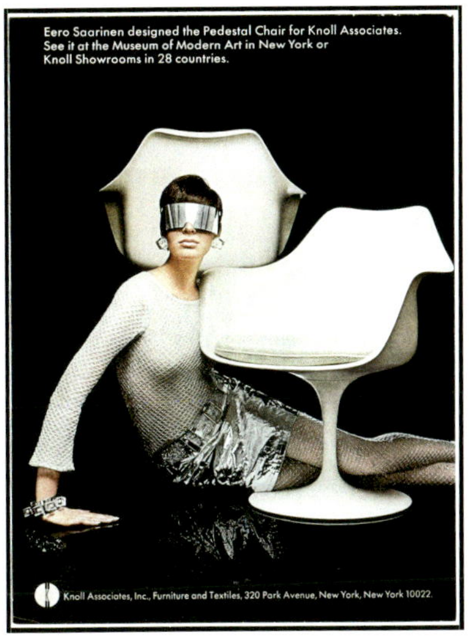

history to bring to light the archetypes and structures proposed by nature. Lovegrove takes a keen interest in paleontology, in the discovery of evolutionary models of living organisms linked to biotope mutations, but also in tribal arts, in the variety of the simplest tools, all the objects betraying behind their function an intelligence of space and environment. These perpetual returns to the original and this look back at the genesis of forms are the fulcrum of the critical analysis, not to say the *deconstruction*, that is preliminary to all his creation.

Glass sponge's lattice-like internal skeleton
Made up of silica, measuring up to 25 cm long

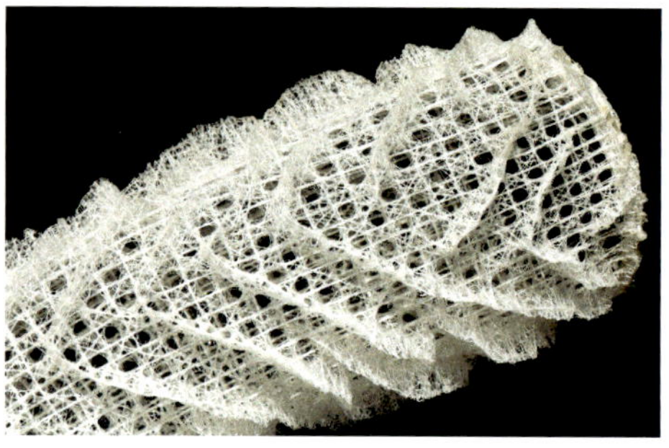

It all begins with the drawings with which he fills his notebooks, like the one titled *Primative Modernism*, in which the density of lines yields form types. This involves a patient labor of appropriation for the designer. He advances like a morphologist fascinated by Constantin Brancusi, Henry Moore, Isamu Noguchi, and all those who have revived transhistoric forms from the depths of human cognition. Turning a car into a simple elliptical form (*Instinctive Override*, 2012) entails going beyond technological complexity to rediscover the emotion, a memory trace that immediately touches our intuition. Whereas many of his objects, like the New Wave Flower Pot (2005), the Muon speakers (2007), the Istanbul Sanitary Ware Collection (2004), and the Eco Book Ends (2011), have this force given by the primal form, other pieces, infused with references and tributes to an ideal family of architects and designers, are embedded in a hypothetical—almost Darwinian—evolution of design history. In the wake of the commission for the Emma Office System (1994–2000) by Herman Miller (Curved Space System, 1972), intended to recompose Robert Propst's Action Office system (1964), Lovegrove understood that technological mutations, the development of new materials, other sociopolitical contexts, and other environmental constraints could have made many now iconic objects evolve in the way natural species do. With this in mind, he revisited George Nelson's Swag Leg (1958) to create the Geo Office Chair (1991–1994); Harry Bertoia's Diamond chair (1952), whose trace can be found in the Lovenet-Bank chair (2002); Arne Jacobsen's famous Series 7 chair (1955), which inspired the Orbit Wood stacking chairs (2003); or yet again the famed Frog (1972) and Egli MRD-1 (1986) motorcycles, from which the carbon-fiber *Ridon* (2009) emerged.
All of Lovegrove's work is informed by the necessity for an optimization and a selection based on a naturalist

Henry Moore
Three Reclining Figures
1976
Graphite, pastel, watercolor on blotting paper
26.3 × 21.7 cm
HMF 76 (8)
The Henry Moore Foundation, gift of the artist, 1979

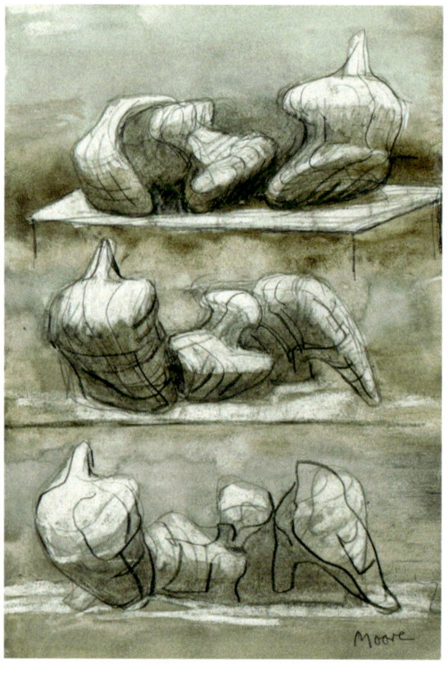

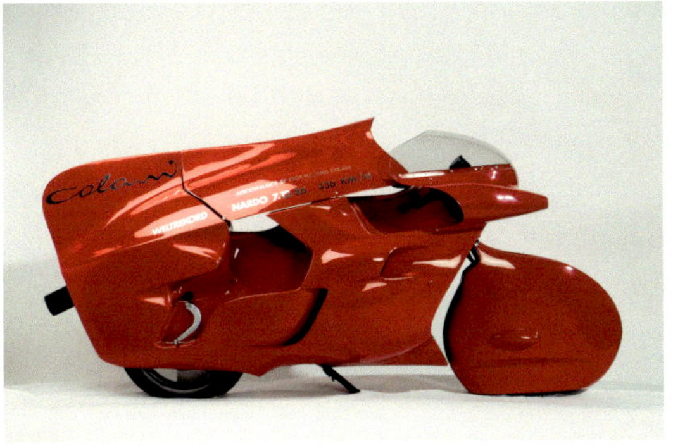

Luigi Colani
Egli MRD-1 motorcycle
1986
Fiberglass-reinforced polyester
130 × 260 × 75 cm
Paris, Mnam-CCI

approach, according to which objects could change in response to a dynamic of their own, their organicity adapting and complexifying with the context and environment. Proud of his filiation with certain designers well-versed in biomorphism, notably Carlo Mollino and Arne Jacobsen, and enthused by the creativity of living organisms that he aspires to reproduce on different scales, Lovegrove sees the organic firstly as a manifestation of life that depends on primal morphogenetic processes, on elementary structuring principles. Beyond his attraction to the evolutionary structure of bone that has led to many of his experiments since the Bone Chair (1994)—from his essential Go Chair (1998–2001), with its structure in magnesium, to the famous *Cellular Automata* series (2008) with its three-dimensional simulation of bone fiber, and the Andromeda (2006–2008) and New Nature (2012) lamps—the designer has been working on modifying forms based on materials, on a sort of metabolism of objects. Known as "Captain Organic" because of the way the curves of his creations form an extension to the body—in the manner of his Eye Digital Camera (1990), the carbon-fiber Biolove bicycle (2003–2011), his Air One armchairs (1999), the BD Love chair (2000), the Brasilia armchair (2003), and his Supernatural chair (2005–2008)—Lovegrove has developed an extensive vocabulary of aero- or hydrodynamic shapes, arising from a process of controlled morphogenesis. This ought not to eclipse what constitutes the heart of his method, namely, his generative, if not genetic, approach to materials. Lovegrove runs through the materials of industrial catalogues to penetrate the secret of their composition and structuration, in order to reinvent the processes of transformation that reified them, conducted at the product state. Attentive to Janine M. Benyus's research on biomimicry, or biomimetics, he creates a new relation to materials in which biomimetics can support sustainable development, notably through the agency of new interfaces between biology and technoscience. This designer of postindustrial economy advocates a reconceptualization of nature through the advent of synthetic biology and the production of new materials, more adapted to reuse, which would thereby minimize waste.

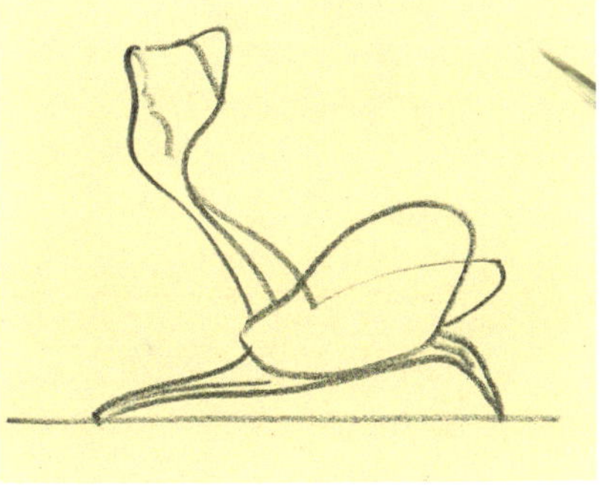

Carlo Mollino
Sketch for his armchair for CADMA,
1947

In fact, many of his projects integrate alternative methods of energy production and management. This is the case for Solar Seed (1994–1995), the first prototype of a solar habitat, and for Solar Car (2008–2009), in a design (produced in collaboration with Cambridge University) that does not consider speed the absolute measure of performance and whose profile remotely recalls the 1964 Bluebird-Proteus CN7. It is also true of the Solar Tree lamp (2006–2007), a new type of hybrid street lighting, and Car on a Stick (2002–2007), a self-driving car that, when not in use as a vehicle, rises to provide street lighting and free up space on the ground. This quest for the complete autonomy of the technological object is found again in Alpine Capsule (2005–2008), a mercury-droplet shaped hotel room set on the mountain top and connected to a solar plant, or in Bioyot (2011), a yacht that is also an insular territory.

Can design, architecture, and engineering create eco-friendly, sustainable entities, simplify to the extreme the conception of structures, limit the number of elements, and find in the continuity and sobriety of forms an economy, an optimization of production costs and conditions? Most often reduced to an envelope, to a simple surface, a volume, a bubble, Lovegrove's creations aspire to a perfect balance between matter and form. This is why the mercury droplet (quicksilver being the only liquid metal, at once dense and mobile) is both the material and the emblem of the Mercury suspension lamp (2009), among others. Lovegrove's obsession with materials and their exploitation, be it copper (Cetacea lounge chair, 2000), magnesium, titanium, carbon (Gingko table, 2003–2006; GT110 suitcase for Globe-Trotter, 2006–2007; Moot chair, 2012–2013), or methacrylate (Cosmic Leaf, 2009–2010; Cosmic Angel, 2009–2011), prefigures the use of new bioengineered fibers, silicon dioxide, and aerogels, bio-based materials, and new polymers that will replace those derived from fossil sources.

James Watson (b. 1928) and Francis Crick (1916–2004) with their DNA model in 1953

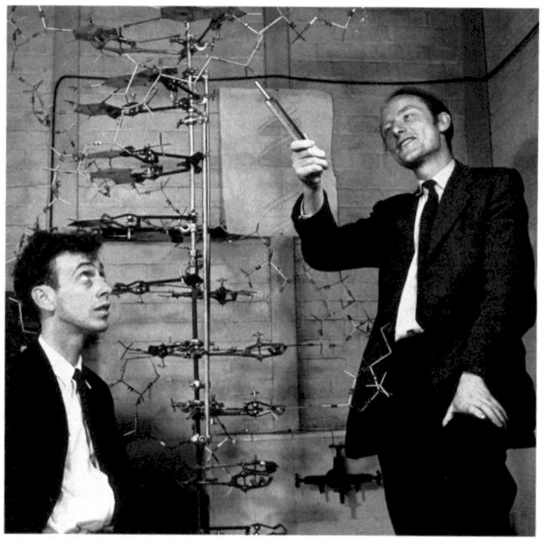

As our relationship to materials changes, so does our understanding of the intelligence of nature, in a world that can no longer be anthropocentric, as the Aristotelian tradition of separation between *physis* and *technē* conceived of it. Lovegrove is intent on establishing a new continuity between nature and technology that is not simply a technological simulation of natural processes. Synthetic biology gives rise to the prospect of a synthetic design, a naturalization of artifacts permeable to the economy of other ecosystems. Such a naturalization—what Lovegrove calls "the nature of naturalness"—invites us to devise another *episteme* and to rethink procedures of industrial production as a whole. This approach takes us back to the origins of humanity, to prehistorical times when the bonds with nature, matter, and the environment of humans were developed, to a period when a primitive relationship to the organic was constructed that became embedded, little by little, in our genes, an *organic essentialism*—one of the designer's key concepts—that underpins our relationship to Earth and to life (DNA staircase, 2005). Seeing himself as an artisan crafting materials and forms, Lovegrove relies on this connection with the earth and with the organic. In it he sees the prospects of a therapy that could purge us of the never-ending mediations of our industrialized world. Design would thus have a cathartic function, liberating us from our "comfort zones" and enabling us to rediscover the primitive interactions between our tools and the effects they bring about. This calls to mind what Henri Focillon wrote in his 1934 essay "In Praise of Hands" (1934): "The artist carving wood, hammering metal, kneading clay, chiseling a block of stone keeps alive for us man's own dim past, something without which we could not exist. Is it not admirable to find living among us in the machine age this determined survivor of the 'hand age'?"[2]

A block of aerogel presented by Peter Tsou, STARDUST project coordinator at the Jet Propulsion Laboratory

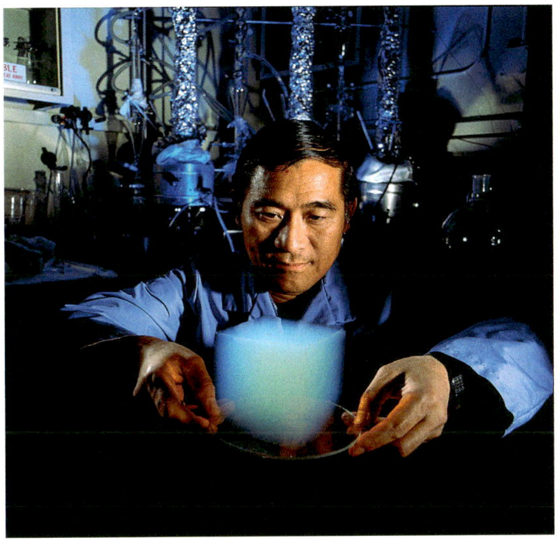

Notes

1. Alfred North Whitehead, *The Concept of Nature* (Cambridge, UK, 1964), p. 14.

2. Henri Focillon, *The Life of Forms in Art*, trans. George Kubler (New York, 1989), p. 169.

Morphogenesis, Magic, and Empathy: Ross Lovegrove's "Animate Forms"

Marie-Ange Brayer

Ross Lovegrove works in his office as in a Renaissance *studiolo*, surrounded by his own creations in conversation with natural specimens, artifacts, archaic objects, and artworks, in the same way one would find *mirabilia* in a cabinet of curiosities juxtaposed with *naturalia* and *scientifica*: objects of wonder alongside natural and scientific objects. In *The Lure of Antiquity and the Cult of the Machine: The Kunstkammer and the Evolution of Nature, Art, and Technology,*[1] art historian Horst Bredekamp analyzes the way in which cabinets of curiosities in the sixteenth and seventeenth centuries "favored an early imaginary of nature's evolutionary development."[2] Lovegrove emphasizes an overall vision of design that finds its origin in the evolutionary processes of nature. His creative approach is marked by a *physis* that crosses the natural and the artificial, drawing from both the natural sciences and computational architecture. The genealogy that he draws up of his own projects calls to mind the tree-of-life diagrams of the German biologist and philosopher Ernst Heinrich Haeckel (1834–1919) as well as Charles Darwin's theories. But, according to Bredekamp, Darwin thought of evolution less as tree-like than in the form of "swarms of non-linear points."[3] Similarly, the genealogy of works by Lovegrove summons a temporal dynamic in which works are collected "around certain points,"[4] around singularities, within morphological groups with their own formal mutations.

Levinus Vincent
A Dutch cabinet of curiosities
Illustration from *Wondertooneel der Nature* (Amsterdam, 1706)
BNU collection, held at the University of Strasbourg

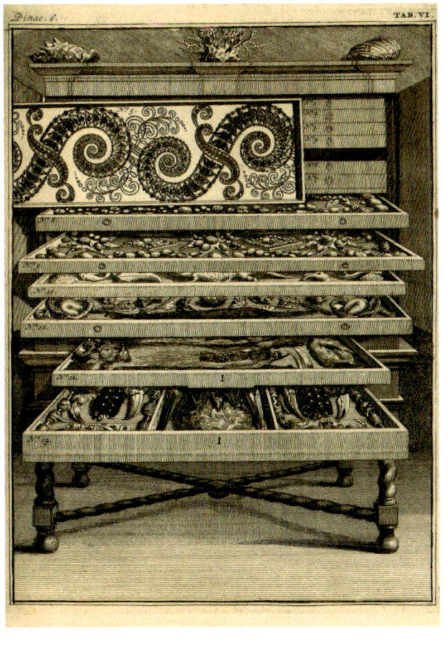

Scientia naturalis and Morphogenesis

In *Darwins Korallen* (Darwin's corals, 2005), Bredekamp speaks of Darwin's world as animated by iconology,[5] a world in which art and biology are in resonance. When the naturalist examined corals in *The Origins of Species* (1859), he was more interested in them as evolving structures than as organisms. He undertook a taxonomic and genetic exploration of corals in *The Structure and Distribution of Coral Reefs, Being the First Part of the*

Geology of the Voyage of the Beagle (1842). During the same period, Jules Michelet discussed the organic constructions of corals in *La Mer* (*The Sea*, 1860) as resulting more from their own morphological mutation than from the influence of external factors. Already in the eighteenth century, Johann Wolfgang von Goethe had studied the dynamic morphology of plants in *Versuch die Metamorphose der Pflanzen zu erklären* (*Metamorphosis of Plants*, 1790). In the meantime, Haeckel was investigating the genesis of life as an interaction between organisms (*Generelle Morphologie der Organismen* [General morphology of organisms], 1866; *Kunstformen der Natur* [*Art Forms in Nature,* 1899]). In 1866 he forged the term "ecology" from the Greek word *oikos* to refer to the interaction between a type of organism and its environment. In 1917 the Scottish biologist and mathematician D'Arcy Wentworth Thompson's discussion of the morphogenesis of living organisms in *On Growth and Form* left a lasting mark on biological thinking.

These studies of life had a considerable impact on the historical avant-gardes and on such movements as Expressionism, and they affected our understanding of the notion of "structure" in the fields of art and architecture. In the 1930s, the architect and engineer Robert Le Ricolais, influenced by the work on radiolarians by both Haeckel (*Radiolaria,* 1862) and Thompson, studied automorphism, the characteristic growth of forms, with particular attention to defining the "architectural" nature of these micro-organisms with complex polyhedral forms. "Shells, radiolarians, and crystals are all examples of veils, networks, and skeletal structures."[6] From the crystalline aggregates to spatial agglomerations, architectural structure is understood through the principles of morphogenesis and self-organization in living organisms. Le Ricolais explored three-dimensional crystalline networks, seen as inhabitable structures (*Essai sur les systèmes réticulés à 3 dimensions* [Essay on three-dimensional networked systems], 1940). His soap-film experiments in the 1930s paved the way for the use of spatial, or three-dimensional, structures in experimental architecture of the 1950s and 1960s. With "form no longer dependent on a geometrization of space,"[7] the principal of generativity can take hold.

In *Theory of Self Reproducing Automata* (1945), the Hungarian-American mathematician John

Ernst Haeckel
Acanthopracta
Lithograph by Adolf Glitsch after a sketch by E. Haeckel, *Kunstformen der Natur* (Leipzig and Vienna, 1904)
Library of Congress Prints and Photographs Division
Washington, DC

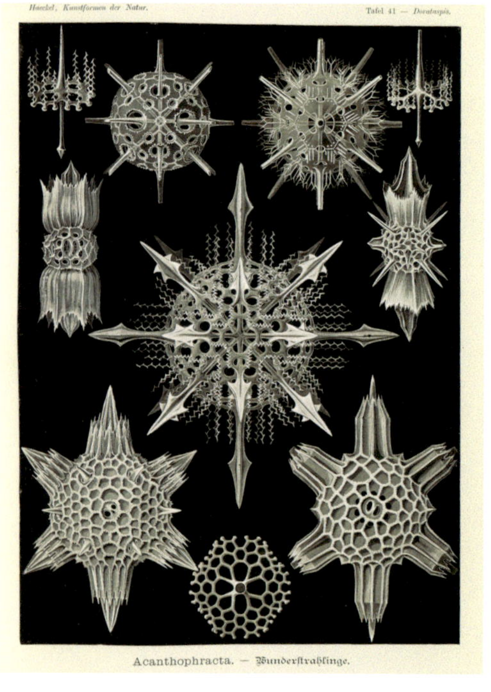

von Neumann studied cellular automata—mathematical models that self-replicate based on cell interactions that will be the basis for simulation software. "It was starting in the 1950s that computer simulation came to be used to model morphogenesis."[8] The French mathematician René Thom emphasized the connections between mathematics and nature through the concepts of "morphogenesis" (*Stabilité structurelle et morphogenèse* [*Structural Stability and Morphogenesis*], 1972) and "semio physics." In 1968 the Hungarian biologist Aristid Lindenmayer defined principles for modeling plant growth that were to influence the language of programming for computer-aided design (CAD). In the 1950s and 1960s, the geodesic domes designed by Richard Buckminster Fuller relied on a mathematical approach to nature and the study of radiolarians, as did the projects of the German architect Frei Otto, who investigated light structures in the 1950s. His research into the principle of self-generation in nature also drew inspiration from the functioning of single-cell organisms that Haeckel and Thompson examined. For his architecture projects, Frei Otto studied the ontogenetic development of living organisms and their principles of growth for purposes of structural optimization.

Digital Ecosystem

With the spread of digital technologies in the 1990s, many architects discovered in turn the convergence between these studies of morphogenesis and digital simulation processes to recreate the principles of living organisms. The digital revolution developed in interaction with life sciences, synthetic biology, thermodynamics, physics, artificial intelligence, and the cognitive sciences. Using CAD/CAM (computer-aided design and manufacture) software, architects can explore the implications of a systemic approach to biological life, studying its possibilities of replication, self-organization, and emergence, which intersect the computational "ecosystem." The American architect and theorist Greg Lynn was one of the first to explore these convergences in his book *Animate Form* (1999), which draws on the vitalist philosophies of Henri Bergson and Gilles Deleuze. By the late 1990s, the evolutionary process of animate forms that Lynn introduced set in motion a radical change in the field of CAD. In his book, Lynn talks of the influence of Thompson and the German art historian Wilhelm Worringer, whose concept of "animation of the inorganic," developed in *Abstraktion und Einfühlung* (*Abstraction and Empathy,* 1907),[9] inspired Lynn's own concept of animate form. The "blob," as defined by Lynn, behaves like an organic entity that self-regulates in response to actions and forces present in the environment. It lends itself to a generative exploration of form, anticipating the multi-agent algorithmic systems and their morphogenetic dimension. Architecture is thus understood as "a system of dynamic organizations,"[10] in which "forms evolve and interact with each other."[11] In 1990, when Lovegrove opened his Studio X in the Notting Hill district of London, he was already a recognized industrial designer, known in particular for his design of the Walkman for Sony and his work on Apple's iMac computer. In 1994, after a stint at Knoll International in Paris from 1984 to 1987, where he created the Alessandri Office System, he started work on an integrated office system for Herman Miller using polymers molded in the form of honeycombs. This was followed by his design for a new product that radically renewed the language of the design object and its use: the Eye Digital Camera for Olympus (1990). Shedding the traditional hard angular metal camera body, the Eye Digital Camera made of soft synthetic rubber and boasts tactile properties and a biomorphic form that adapts to the human body. Lovegrove cites the membrane surface designed by Mario Bellini (1935) for Olivetti calculators and computers (*Divisumma 8,* 1972) as having inspired him for the Eye Digital Camera. The object here is no longer simply a useful form. For Lovegrove, its "physicality"—the complex of functional and sensorial parameters—transforms it into a prosthetics, an extension of the body. The fluidity of such objects produces an unprecedented sensorial experience, akin to the kind of experience that a designer like Ettore Sottsass aspired to achieve with the archetypal charge of his *Ceramics of Darkness* (1963), intended to put us in touch with a primordial cosmic dimension. Beyond their function in everyday life, objects, according to Lovegrove, should connect us to a symbolic level. "When I see such artifacts in a museum or when I hold them in my hands, they seem to transmit a form of energy to me," he says.[12] And this is certainly the case for the Eye Digital Camera. In its design, the Eye Digital Camera thus resembles the soap-film models of the architect and engineer Le Ricolais and later those of Frei Otto to create minimal surfaces. This

Frei Otto
Project for minimal surface over a meeting point for students, made of soap film,
1988

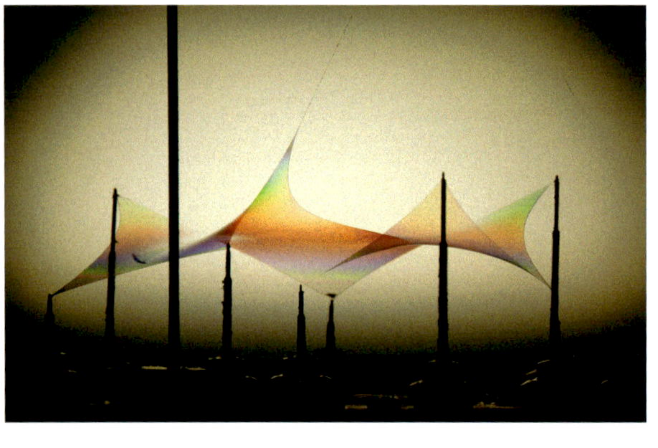

"bioanatomical" object attests to Lovegrove's intelligence in the use of materials and new industrial processes. In designs ranging from everyday objects to cars, aviation, and architecture, he has developed his concept of "DNA" (design, nature, art), creating a powerful link between digital technologies, the science of materials, and organic forms. His approach owes something, no doubt, to the biodynamic forms of the German designer Luigi Colani, but it was the digital revolution that allowed Lovegrove to go farther in his aesthetic quest, beyond the notion of form to open onto an evolutionary dimension.

From his very first projects until today, Lovegrove has demonstrated again and again that he is perhaps the single designer who best understands the radical upheaval generated by digital technologies. He is practically the only one of his generation to have developed a total complicity with digital experimentation in architecture, sustained by his intellectual proximity and friendship with such architects as Greg Lynn and Zaha Hadid. Indeed, Lovegrove has embraced the generative design concepts in architectural projects made possible by new parametric modeling software. He is also the only designer to take part in forward-looking exhibitions with respect to digital technology—*Latent Utopias* (Landesmuseum Joanneum, Graz, 2003, curators: Hadid and Patrick Schumacher), *Naturaliser l'architecture* (ArchiLab, Orléans, 2013, curators: M.-A. Brayer and F. Migayrou)—and to publish in specialized architecture journals.[13] His exploration of natural sciences is inseparable from the emergence of a computational field whose reach now extends to all scientific and industrial sectors, and whose aesthetic and structural potentials are the focus of Lovegrove's interest. Hence, it can be said that he anticipates in the field of design "the new generation of architects who have . . . blurred the boundaries between architecture, physics, and biology to invent patterns of interactions between forms and materials, between the mineral and the organic, between construction process and growth."[14]

Zaha Hadid Architects
Kartal Pendik Masterplan, Istanbul
2006

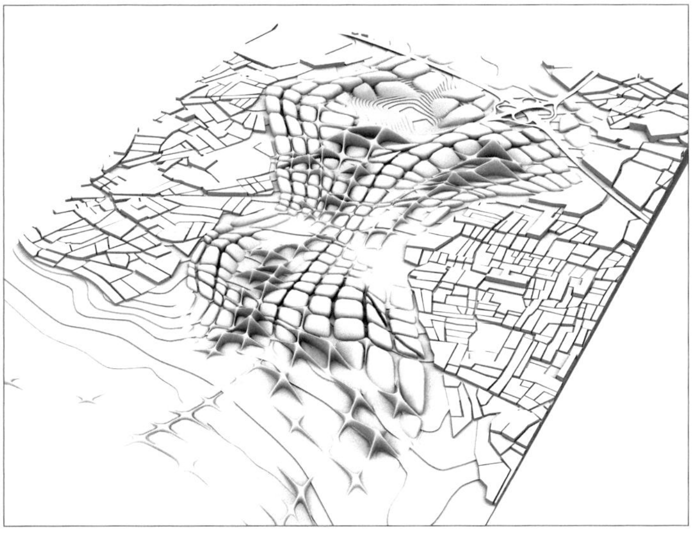

From Hylomorphism to a Haptic Dimension

Lovegrove develops a novel process for the design and manufacture of each object. The water bottle for Tŷ Nant (1999–2001), whose shape is modeled on the rippling movements of water, was one of the very first digitally designed industrial products. Developed from nontoxic polymers, it embodies the "DNA" (design, nature, art) concept he advocates. Lovegrove drew on a variety of sources to express the morphological convolutions of water, including Leonardo da Vinci's water studies, the sculptor Henry Moore's drawings of animal bones and skulls, and the movement of the fashion designer Issey Miyake's fabrics. Black-and-white photos, preliminary sketches, and digitally modeled 3-D sculptures punctuated a long, complex period of research spanning more than a year. Just as the builders of the Nasrid palaces of the Alhambra in Grenada from the eighth to the fourteenth century set out to create geometric patterns from the flow of water through the canals that irrigated the gardens, Lovegrove undertook to capture the essence of water, express its fluidity, and its impermanence—in his own words, to "put a skin on water itself." The Tŷ Nant bottle does not *represent* the movement of water; it incorporates water's cinematic nature, its specific energy, and the tension between opposing forces that run through it. At once optic and haptic, the perception of the object becomes an atmospheric field.

Greg Lynn
FORM
Embryological House
1999
Exploded kit of parts for House A and House C

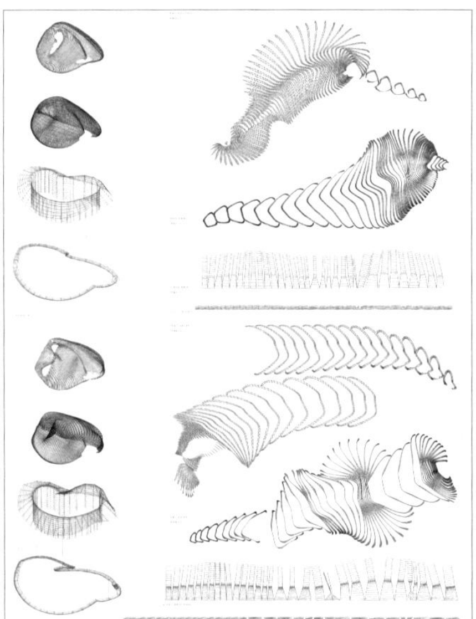

Tools of parametric design and the *processes* of digital modeling have enabled Lovegrove to express the unrepresentable, by thinking less in terms of form than a form-generating process. The Tŷ Nant bottle is the first industrial design object to use the algorithmic languages developed for architecture, that is, the CAD systems, based on NURBS (Non-uniform rational B-spline), that such architects as Frank Gehry and UNStudio were using in the same period to "produce singularities on a continuous surface."[15] Enveloping and flexible, B-spline surfaces generated in this way develop beyond basic geometric shapes. Lynn stresses the fact that digital shapes are not organic, and that their organic appearance "is the result of organizational principles based on differentials."[16] As the Italian historian and architectural theorist Mario Carpo observes, parametric design does not consist in producing "an object but an algorithm—a parametric function which may determine an infinite variety of objects, all different . . . yet all similar."[17] The emergence of digital tools led to a radical paradigm change in the field of architecture, breathing life into new concepts (fluidity, continuum, hyperspace, data-scapes, etc.) that challenge all formal assumptions. Lovegrove's investigations into the dematerialization and parametric singularization of forms has found an echo in such contemporary architectural projects as HtwoOexpo, the water pavilion (1993–1997), built by Lars Spuybroek (NOX) in the Netherlands: an interactive architecture that puts water in all its forms into play, by exploring principles of liquefaction, immersion, and the fluidity of data. By the early 1990s, proclaiming that form is "grown" rather than "constructed," Lovegrove's work was instrumental in radically renewing the concept of the design object. His explorations at the time intersected the new computational architectural field that the architecture historian and critic Detlef Mertins termed "bioconstructivism."[18] In studying the principles governing the growth of natural forms, Lovegrove aspires to design structures that are at once simple and sophisticated in innovative lightweight materials. Biomimetics and minimal energy consumption are at the heart of investigations that would influence designers of younger generations, such as the Japanese Tokujin Yoshioka and the Dutch Joris Laarman. The Crystal Chair presented by Tokujin Yoshioka in 2007–2008 at *Second Nature,* an exhibition at the museum 21_21 Design Sight in Tokyo, in which Lovegrove participated, manifests a similar approach. Indeed, this chair is quite literally grown rather than constructed, as part of the natural growth process of crystals. Similarly, Lovegrove's work on aluminum and carbon had an impact on the optical effects of Tokujin's objects in molten glass (such as the Water Block bench, 2002). "It is not soft curves that characterize his design method . . . [but] the interactive relationship between humanity, nature and technology,"[19] says the Japanese designer about Lovegrove. Form, for the latter, is not an end but rather the genetic unfolding of an organic process of growth. His series of "liquid" objects in carbon fiber or aluminum—Liquid Carbon (2007) or Liquid Shelves (2011)—conjure an "elastic" space-time in which forms seem suspended, undecided, in a state of evolution, neither solid nor liquid. They express the tension between the intrinsic and extrinsic forces of the materials. The animation of the inorganic, characteristic of Lovegrove's

NOX/Lars Spuybroek
FreshH2O, the water pavilion
1993–1997
HtwoOexpo,
Neeltje Jans,
Netherlands

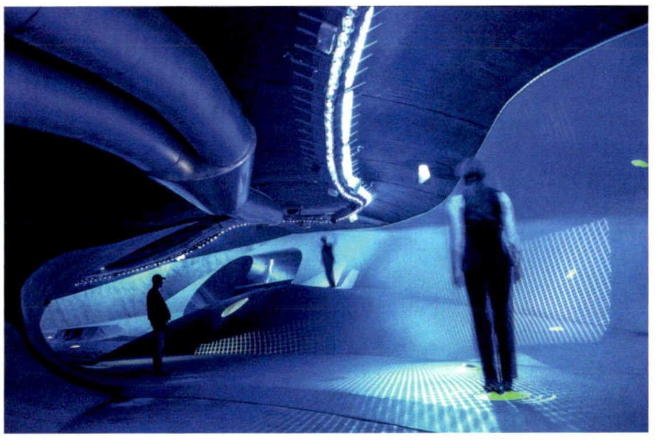

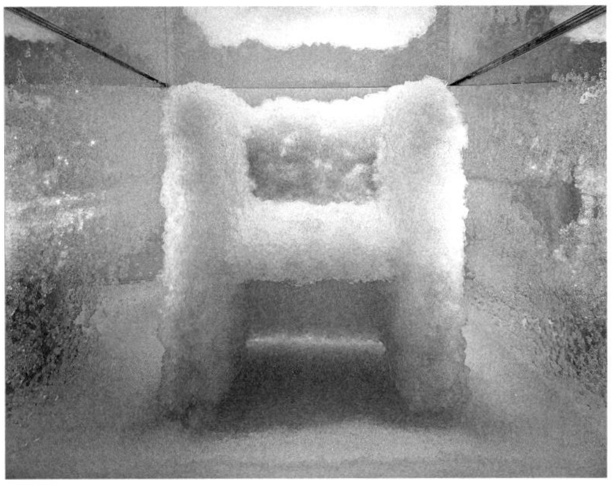

Tokujin Yoshioka
Venus chair
2008
Natural crystals
Variable dimensions

method, is related to the work of Haeckel, who examined (in *Kristallseelen: Studien über das anorganische Leben* [Crystal souls: Studies of inorganic life], 1917) the existence of "biocrystals," "a mixture of mineral substances . . . with organic plasma," which had "the ability to form a skin through which they appear to breath like living organisms."[20] It also echoes the research of crystallographer Otto Lehmann in the early twentieth century on the "spectacular" plastic qualities of "liquid crystals."[21] These designs by Lovegrove hover between liquid and solid like a mutant morphology, neither altogether matter nor altogether form.

If the Tŷ Nant bottle can be described as "a sculpture of light and shadow," it also evinces the emotional thrust of his designs. Paola Antonelli, senior curator at the Museum of Modern Art in New York, uses the expression "compassionate innovation"[22] to describe the method applied by the designer to establish empathy between the object and the person perceiving it. Deploring the way in which "the process of creation in the industrial world is becoming more and more detached from the origin of materials and processes," Lovegrove has called on designers to reestablish a connection with "the deep sensorial origins of how we perceive the world around us."[23] For him, design is less a matter of producing objects than of triggering an empathetic relationship, enhanced by a synesthetic dimension, in the sense that the subject is incorporated in the object's immaterial aura. This is the case for the endlessly changing orbs created by the movements of the Tŷ Nant bottle or the spatial vibrations of his Liquid Shelves, whose intermediary state between liquid and solid, animate and inanimate, manifests a new "ecophysics of the artifact,"[24] between nature and artifice. The shapes become animate, harking back to a primordial consciousness of the material and to the magic and symbolic dimension of early artifacts. According to Sottsass, design begins where "magical processes begin,"[25] and it is unfortunate that we can no longer understand them or know how to use them, that while prehistoric arrows still work, we are not incapable of interpreting their graphic signs. In much the same way, Lovegrove would say that technical objects without empathy or magic force have no reason for being.

Organic Essentialism

Lovegrove forged the expression "organic essentialism" to shift the emphasis from a form that calls to mind that of an organism to the generative dimension that is decisive in the design and development of all living organisms. His Go Chair (1998–2001), the first chair in injection-molded magnesium, symbolizes this organic essentialism in a "fat-free" object that is literally reduced to its force lines. The chair, the mass, and the volume have given way to the skeleton, to the bone structure. In the wake of Alberto Meda's carbon fiber Light Light Chair (1987), which reduces the dynamic form to a single skin-like surface, the Go Chair, with its perforated surface, recalls bone cavities and bone growth. In 2006, the Dutch designer Joris Laarman's Bone Chair also modeled itself on the principle of bone growth in living organisms, while drawing inspiration as well from the work on plant life by the German physicist Claus Mattheck, the bionic pioneer (*Design in Nature: Learning from Trees,* 1998). Optimization algorithms that generate 3-D simulation of natural growth processes and the distribution of forces have made it possible to calculate weight distribution to enhance structural performance. Lovegrove's objects are "not so much designed as evolved with the feel of a biological organism that grows from intrinsic and extrinsic forces, rather than being mechanically constructed."[26] This generative approach is made possible by the digital ecosystem that serves to "animate" the inanimate object. If Lovegrove, in his search for essential forms, readily sees himself as heir to such creators as Charles and Ray Eames, Arne Jacobsen, Eero Saarinen, and Isamu Noguchi, for whom the organic wells up out of the inorganic, the term "organic" goes beyond the notion of biomorphism—the imitation of natural forms that prevailed in postwar organic design, as it was presented, for instance, in the *Organic Design in Home Furnishings* exhibition at the Museum of Modern Art in New York in 1941. Henceforth,

Charles and Ray Eames
Molded Plywood Sculpture Experiment
1942
Eames Office, LLC

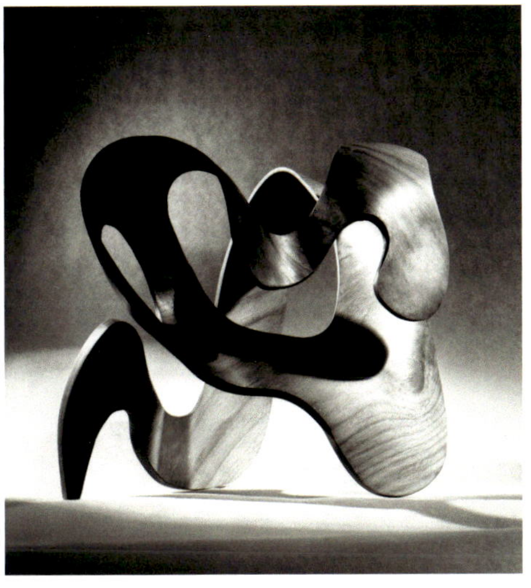

Charles and Ray Eames
Chaise Longue, Full Scale Model
1948
Hard rubber foam, plastic, wood, metal
82.5 × 149.9 × 87 cm
Museum of Modern Art, New York
Gift of the designers
SC106.1973

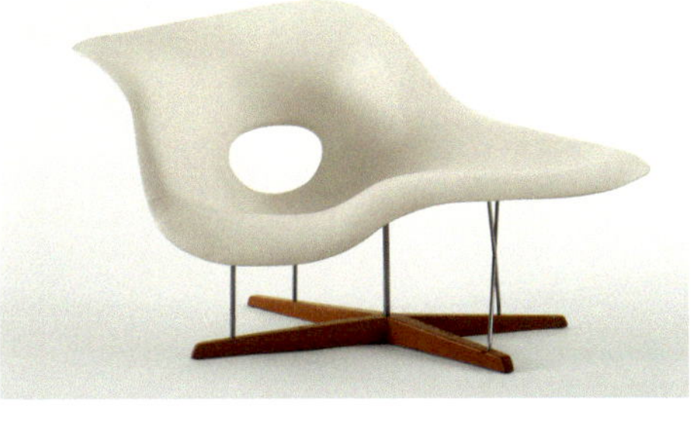

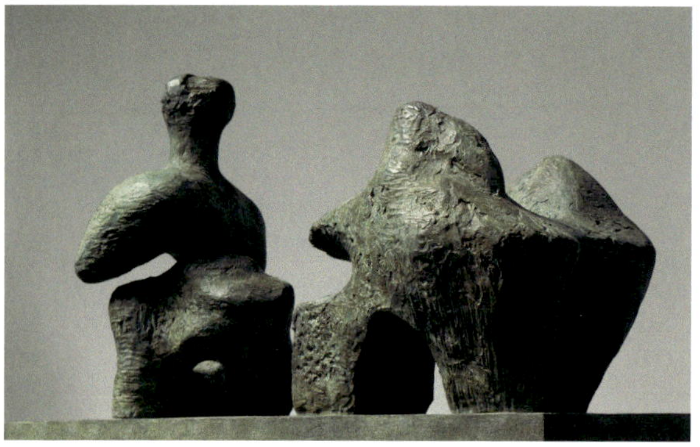

Henry Moore
Two Piece Reclining Figure No. 4
1961
Bronze
L.: ca. 109 cm
(LH 479)
The Henry Moore Foundation, gift of the artist, 1977

the notion of organic no longer refers to curvilinear forms but to the *convergence* between the object's structural materiality and natural forms in constant metamorphosis. The inflection of the line expresses the thrusting forces intrinsic to the object. To the architect Kengo Kuma, a "thing that is organic is different from a thing that is simply natural or made of natural materials. A thing that is organic must possess the generative dynamism characteristic of living things."[27] "The term 'organic essentialism,'" Lovegrove explains, "is only one way of expressing an approach to the creation of forms in space." What he means by "organic" are "structures that are fluid and uninterrupted" and by "essentialism" he designates "the search for the natural essence of a physical object in terms of its inherent combination of materials and form"[28] to achieve the "evolutionary morphology" of the object. His intention is to achieve a convergence between evolution, material, form, and engineering, rooted in a global conception of art and design.

Lovegrove also draws on a convergence between art and design from a period when design objects and artistic objects participated in a single aesthetic register. Ludwig Mies van der Rohe thus conceived his furniture for the Tugendhat House (1930) in Brno (former Czechoslovakia) in the same vein as a Wilhelm Lehmbruck sculpture. Likewise, in the postwar period, a similar spatial conception was applied to design objects and sculptures that "occupy a similar space."[29] After the 1941 *Organic Design in Home Furnishings* exhibition, the Charles and Ray Eames Chaise Longue (1948) was placed in the garden of the New York Museum of Modern Art opposite Gaston Lachaise's *Floating Figure* (1927) sculpture, and the Lounge Chair with Ottoman (1956) by Charles Eames was compared to Henry Moore's *Reclining Figure* (1951). The design object has in fact the same type of relationship with its environment as sculpture, and the Eames chairs can be said to be "sculpting space."[30]

Henry Moore
Reclining Figure
1951
Plaster and string
105.4 × 227.3 × 89.2 cm
Tate, London

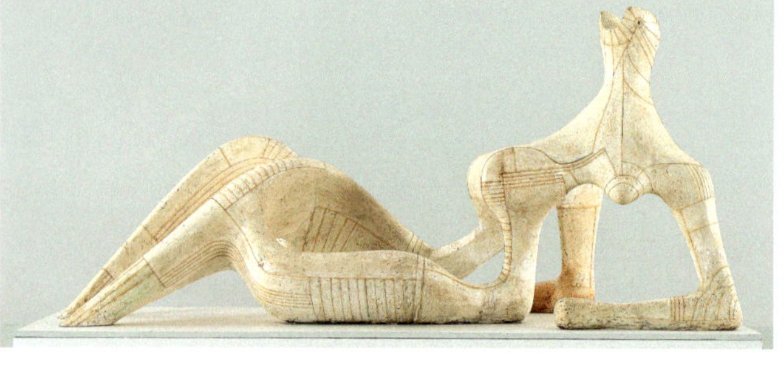

There is no doubt that the observation of nature is a source of inspiration for Lovegrove, but so are Hans Arp's organic forms, Constantin Brancusi's tectonic expressions, Henry Moore's figures with their large gaps, Francis Bacon's convulsive paintings, Antony Gormley's anthropomorphic sculpture, Tony Cragg's "kinetic" pieces, or Anish Kapoor's installations. Lovegrove's Supernatural chair (2005–2008) echoes pieces like *The Helmet* (1939–1940) by Moore, who was himself close to architects and designers, and who stressed the "remarkable structural strength" of bones[31] and the voids in his sculptures. The two-meter-tall aluminum speakers Muon (2005–2007), sound sculptures with an architechtonic look designed to create what Lovegrove called a "liquid experience," are inspired by the abstract sculptures of Barbara Hepworth and Moore, and by Brancusi's *Endless Column* (1937).

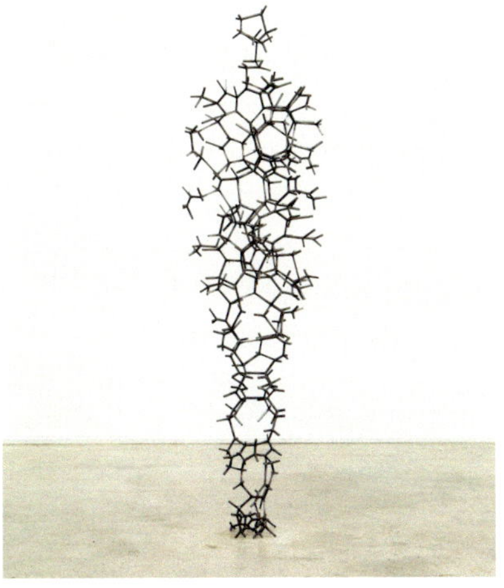

Antony Gormley
Aperture
2009
6-mm-square section mild steel bar
188 × 51 × 34 cm
White Gallery, London

Lovegrove's carbon fiber kinetic sculpture *Ridon* (2005) is a tribute to Isamu Noguchi's *Black Sun* (1969) and *Energy Void* (1971) as well as to the Futurist sculptures of Umberto Boccioni (*Unique Forms of Continuity in Space,* 1913), and he has used the expression "functional sculpture" to describe his Liquid series of shelves, benches, and tables.

Lovegrove sees art as an "invisible force" that can animate an inanimate object. Animation, according to Lynn, "implies evolution of a form and its shaping forces," at the time of conception.[32] This kinetic approach has long been applied in aviation and automobile manufacturing. It is inherent to the generative or computational design in architecture. Lovegrove makes use of motion vectors to accentuate the performative dimension of the form. His project for the Renault Twin'Z (2012–2013), an electric concept car manufactured using state-of-the-art digital modeling technologies, evidences this approach of animating the inanimate. The designer was not involved in the overall morphology of the car, but he endowed it with a kinetic dimension on the glass roof, using a LED system to create light patterns that turn light into energy and set the car virtually in motion.

Instinctive Override, an installation presented at the 2012 Biennale of Kortrijk, also takes the automobile as a starting point. Here reduced to a basic form reminiscent of a drop of water or a cetacean, its oval surface is swept by a stream of light filaments from

Tony Cragg
Versus
2011
Bronze
150 × 160 × 55 cm
No. 13572
Marian Goodman Gallery

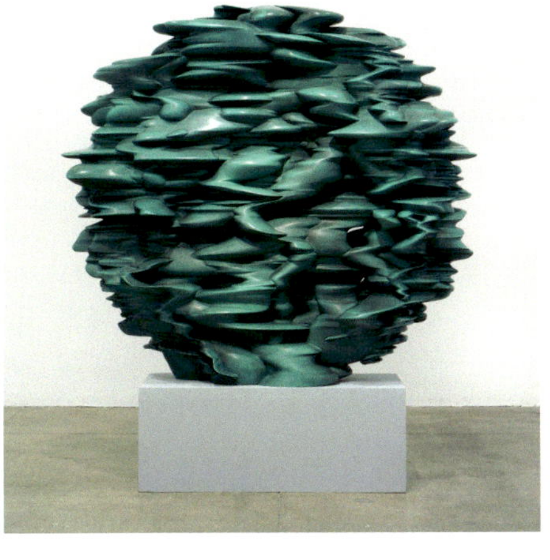

the video projection (made with the participation of architect Alisa Andrasek for the digital design). This "drop of water that travels,"[33] like the pelagic motion of an object in a plasmatic digital environment, seems to hark back to "the primitive beginnings of architecture," that Frank Gehry believed came from "zoomorphic yearnings and skeletal images."[34]

Mutability of Forms

For Lovegrove, the design object is not the representation of a form but rather its morphogenetic expression, where art and science converge. Its "organicity" emanates from the integration of a principle of growth. A case in point is the Gingko table (2003–2006), modeled on the dynamic principle of a ginkgo biloba leaf, which results from the interaction of two carbon-fiber leaves folded and unfolded to distribute the material's structural forces. In extremely lightweight carbon fiber, like the GT110 suitcase for Globe-Trotter (2006–2007), it makes use of the same advanced composite technology being developed in automotive and aviation industries. With a texture resembling a ridged metallic skin, it seems to float like a leaf in the air, escaping the force of gravity.

The stackable Diatom chair (2014, Moroso) in pressed aluminum, also applies automobile technologies. This object is designed and engineered like the body of a car or the wing of an aircraft, but it also owes much, according to Lovegrove, to the structure of diatoms, primordial single-cell organisms, from which he drew inspiration to ensure the product's lightness and resistance. Parametric design tools allowed him to reduce the chair's thickness and stiffen the material through the surface curvature and the pressed pattern of radiating dots. The pattern of slots yields different densities while also providing ventilation. The Diatom chair thus incorporates an evolutionary morphology that explores the boundaries between natural shapes, materials, and technologies.

In the way he exploits the intrinsic properties of the material, Lovegrove's approach resembles that of architects in the digital age. It is the material itself, its physicality, that generates the form, as the mechanical gives way to the performative. The new computational artifact has its origin in its interaction with environmental conditions and is defined through its material behavior. "As opposed to merely translating biomorphic patterns from biology to architecture," writes the architect Achim Menges, "the transfer of knowledge happens on a systemic and performative level . . . Consequently biomimetics is seen as a strategy for architectural and structural design. The goal is to implement a biologically informed material system that draws equally from design principles in nature and from those in fabrication."[35] Materials are the vectors of these formal mutations. Like in the architecture of Frank Gehry, for instance, digital tools and their interconnectivity are in resonance with the object's metabolic dimension. The new computational artifacts (design objects, architectural projects, etc.) surpass formal limits in their metamorphic dimension, originating in their digital materiality, between nature and artifice. Like Gehry, fascinated by the kinetic dynamic that propels forms into a space-time deformation, Lovegrove introduces a morphogenetic dimension into his designs. His luminaries for Artemide elude physicality to compose veritable light environments. The result of algorithmic calculations, Cosmic Leaf, a digital leaf floating in space, is a light-capturing device, with the light diffracting through color filters. In many ways Lovegrove's lights hark back to Bruno Munari's Useless Machines from the 1930s and 1940s—suspended aerial sculptures whose chromatic elements were set in motion by atmospheric changes. The materials here are air and light, and these works lend themselves to a morphogenesis of forms. Their cinematic dimension has a poetic and emotional thrust. According to Munari, "The Useless Machines" should trigger an unstable perception through the dynamic transformation of perpetually developing forms, suspended between two and three dimensions, between the physical world and their virtual projections in space. The dynamic process to which Lovegrove subjects his forms goes hand in hand with the active, empathetic reception of his pieces. The Chlorophilia suspension light (2015), molded from thermoplastic, is designed as an installation of transparent leaves marrying natural and artificial geometries. The optical plays of light in the form itself and in the light patterns cast on the ceiling create an "expansion of space" (Lovegrove). Other designs, like Mercury (2009) and Botanic Rain (2012), constitute installations of changing forms in space, as if animated from the interior and in a state of suspension between crystallization and dematerialization. They bear similarities, according to Lovegrove, to the structure of animals, such as cetaceans (Cetacea lounge chair, 2000; *Instinctive*

Bruno Munari
Macchina Inutile
1949
Metal, Plexiglas, wood, nylon
87 × 30 × 30 cm
Paris, Centre Pompidou, Mnam-CCI

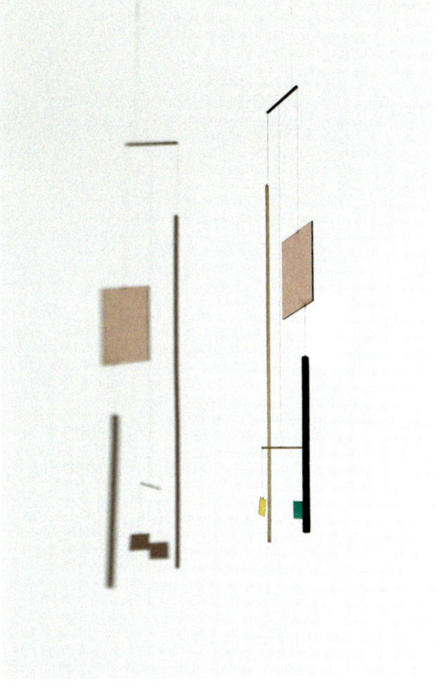

Override, 2012). The spiraling of his Infinite Loop lighting (2014, for Barrisol) relies on the textile's flexibility and stretchability. Using digital modeling, Lovegrove has succeeded in simulating the behavior of the material and making digitally designed structures and forms converge to create lights that appear to be "volumetric surfaces."

Generative Architectures

Lovegrove has long aspired to bridge the gap between the profession of designer and that of architect, where all dimensions of his research on materials and morphogenesis converge, while opening onto a nonlinear world that acknowledges the end of the analogue era. Already in 1997, with the Villa Matoso architecture project in Porto, he explored new properties for ceramics, which could now be done using 3-D printing. This project consisted in "large polymer structures"[36] that worked to reduce the material density of the built environment. Lovegrove studied topographies of growth by way of a subtraction of materials, in something of an "inverted dynamic." Between 2003 and 2005, he designed the DNA staircase, in carbon and glass fiber, rising like a bone structure in the shape of the double helix of DNA. Combining strength and flexibility with an evolutionary dimension, this stairway paves the way to new principles of architectural design. Similarly, the models for *Cellular Automata* (2008) are the fruit of research on growth processes based on scans of human bones. Done using stereolithography, by coring the material, each of the six studies that compose this work was enlarged to make the bone tissue and its cellular matrix visible to the naked eye. Both *Cellular Automata* and the DNA stairway are growing structures inspired by biology and a cellular understanding of the materials. Corolised Chairs (2012) also boast their formal mutability, based on genetic principles. These chairs, meant to evoke prehistoric artifacts, feature a porous architecture obtained by 3-D printing in alumide, a combination of resin and aluminum dust. The "chair," like a living organism, seems to present variations according to the different stages of an evolutionary process. The Corolised Chairs are microarchitectures of biological structures that speak of an invisible order of things.

From this perspective, scale is of little importance, and a bottle of water can become a "fluid architecture." The Lasvit Liquid-Kristal pavilion (2012), a haptic installation at the Triennale Design Museum of Milan, was transformed into an immersive atmospheric environment. The optical distortions created by the undulating glass panels that form the pavilion embodied the "passage from liquid to solid." The glass panels were designed using parametric tools and each was unique. On the Barrisol canvas roof, the light projection of digital forms generated by multi-agent systems, algorithms based on swarm behavior and the interaction of individual agents, was developed by architect Alisa Andrasek using *Cellular Automata* simulation software. These digital generative processes yield new forms of nature, deploying a "synthetic ecology." As a result, the Lasvit LiquidKristal pavilion appears as "overlaps between the organic and the inorganic."[37] Ilabo shoes (for United Nude, 2015), based on a generative design by engineer Arturo Tedeschi, combine digital design with additive manufacturing. Grasshopper's 3-D modeling software was coupled with a 3-D scan of foot morphology, the verticality of which was heightened by a three-dimensional mesh. Advanced 3-D printing technologies allow for a multidimensionality of the form through multi-material printing, the replication of the process of evolutionary growth, and a dimension of adaptability. Objects with new forms and new materialities will soon make their appearance. Lovegrove sees the potentials offered by additive manufacturing as conducive to bringing nature and new production processes closer together. In this way, in the near future, nano deposition methods will make it possible to calculate the precise quantity of material needed. These nonstandard productions rely on "the idiosyncrasies that emerge from craft- and art-based human experiment."[38] The most advanced technologies, according to Lovegrove, will put us back in touch with primordial forms, so that we will rediscover them like early humans with their first artifacts.

Convergences

Lovegrove is one of the rare designers to put biology, anthropology, physics, and ecology at the core of his productions and to champion a humanist view of design within a holistic approach to creation. For him, the designer is the interpreter or "sculptor" of new technologies. His objects factor in contemporary digital mutations while paving the way to new lifestyles characterized by a "sustainable" consciousness and the harmonization of forms with nature. The "convergence" between art, design, technology, and nature is the guiding principle of his search for a new creative paradigm. The idea of "convergence" is important to Lovegrove because it brings art and science together. It is an overall approach to design that plays an essential role in the twenty-first-century renaissance that we are currently experiencing, and that will lead to tangible creative principles that concern us all, wherever we are in the world. "The idea of convergence intimates a time when the merging of all things takes the form of a profound change in the way we see and make the physical world around us," Lovegrove has stated, "Design is a sphere that is continually being reinvented. Because it involves transforming natural resources into useful objects, the designer is at the very heart of today's technological issues that affect not only our own emotional and aesthetic state but also our collective awareness as human beings—a species that is ever rapidly evolving and needs to adapt continuously."[39]

Lovegrove embeds design in an environmental approach that is bound up with a synergy of animate and inanimate forms. By the mid-1990s, he was advocating the use of solar energy, starting with such designs as his Solar Seed nomadic architecture (1994–1995). In 1998 his Solar Bud light for Luceplan became the first ecolabeled product in Europe for its use of photovoltaic technology. Then there was his integration of bamboo in his 2001 design for the Bamboo Bike. In 2006, the Solar Tree for Artemide incorporated photovoltaic cells and ecosmart LED technology in a tree-shaped street lamp. During the same period, he covered the biomorphic and aerodynamic forms of the Swarovski Crystal Aerospace with a "skin" of polycrystalline photovoltaic cells. Lovegrove has pursued this approach with new modes of transportation and habitation (Kyoto concept car, 2006; Car on a Stick, 2002–2007; Alpine Capsule, 2008) that are destined to shape our future. His Generator House (2006–2008), which takes its name from the project designed by Cedric Price in the 1970s for a smart home, responsive to its resident and based on an evolutionary principle,[40] is a house in polymers and lightweight materials meant to enable its inhabitants to live in symbiosis with the natural environment.

The production of objects will be moving from hard to soft, from heavy to light, according to Lovegrove. A new understanding of quantum physics, the development of bioengineering, and the expansion of digital technologies are offering unsuspected creative horizons that will affect as much the industrial as the scientific world." Design as we now know it will be replaced with computational intelligence," the designer writes, "the physicality of the manmade environment around us will begin to grow organically as an integrated and correlated intelligent system."[41] For Lovegrove, every new product should answer the question, "Why should something exist?" While objects have their origin in a form of violence, their production in the future will result from biological processes governed by a global thinking. The twenty-first-century industrial designer will create biological systems rather than physical objects, to engender a world in which technology will keep pace with issues of health, housing, and mobility; where design, bioengineering, and computational intelligence will make art and science indissociable in a return to the primordial object.

Notes

1. Horst Bredekamp, *The Lure of Antiquity and the Cult of the Machine: The Kunstkammer and the Evolution of Nature, Art, and Technology*, trans. Anthony T. Grafton (Princeton, 1995).

2. Horst Bredekamp, *Les Coraux de Darwin: Premiers modèles de l'évolution et tradition de l'histoire naturelle* [2003], trans. Christian Joschke (Paris, 2008), p. 5. (English trans. based on the French trans. from the German—TN.)

3. Ibid., p. 93.

4. Ibid.

5. Ibid., p. 120.

6. David Georges Emmerich, *Le Ricolais: Dossier de l'Atelier de recherche structural: Histoire de l'ARS* (Paris, 1978).

7. Frédéric Migayrou, "Les réalités du possible," in *David Georges Emmerich, architecte-ingénieur: Une utopie rationnelle* (Orléans, 1997), p. 33.

8. Franck Varenne, "Modèles et simulation: Pluriformaliser, simuler, remathématiser," in *Modéliser et simuler: Épistémologies et pratiques de la modélisation et de la simulation*, vol. 1, ed. Franck Varenne and Marc Silberstein (Paris, 2013), p. 307.

9. Wilhelm Worringer, *Abstraction and Empathy: A Contribution to the Psychology of Style,* trans. Michael Bullock (Chicago, 1997). See also Spyros Papapetros, *On the Animation of the Inorganic: Art, Architecture, and the Extension of Life* (Chicago, 2012), p. vii.

10. Greg Lynn, *Animate Form* (Princeton, 1999), p. 10.

11. See Marie-Ange Brayer, "De la nature et de l'artifice: Affects et artefacts dans l'architecture naturalisée," in *Naturaliser l'architecture*, ed. Marie-Ange Brayer and Frédéric Migayrou (Orléans, 2013), pp. 14–37.

12. Ross Lovegrove, in conversation with the author, December 2016.

13. Ross Lovegrove, "Super-Natural: Parametricism in Product Design," in "Parametricism 2.0: Rethinking Architecture's Agenda for the 21st Century," ed. Patrik Schumacher, special issue, *AD Architectural Design* 86 (April 2016), pp. 102–107.

14. Frédéric Migayrou, "Naturaliser l'architecture," in Brayer and Migayrou 2013 (see note 11), p. 59.

15. Lynn 1999 (see note 10), p. 29.

16. Ibid., p. 19.

17. Mario Carpo, *The Alphabet and the Algorithm* (Cambridge, 2011), p. 40.

18. Detlef Mertins, "Bioconstructivism," in *Lars Spuybroek, NOX: Machining Architecture* (London, 2004), pp. 7–16.

19. Tokujin Yoshioka quoted in Ross Lovegrove, *Supernatural: The Work of Ross Lovegrove,* foreword by Paola Antonelli (New York, 2004), p. 26.

20. Papapetros 2012 (see note 9), p. 121.

21. Ibid.

22. Lovegrove 2004 (see note 19), p. 5.

23. Ibid., p. 151.

24. Migayrou 2013 (see note 14), p. 56.

25. Ettore Sottsass, "Design," *Domus* 386 (January 1962).

26. Lovegrove 2004 (see note 19), p. 37.

27. *Studies in Organic,* ed. Kengo Kuma, exh. cat. Gallery MA (Tokyo, 2009).

28. Lovegrove 2004 (see note 19), pp. 151–152.

29. *Figuring Space: Sculpture/Furniture from Mies to Moore*, exh. cat. Henry Moore Institute (Leeds, 2007), p. 5.

30. Joseph Giovannini, "Chairs by the Eames Office: Sculpting Space," in ibid., p. 72.

31. Ibid., p. 63.

32. Lynn 1999 (see note 10), p. 9.

33. Marie-Ange Brayer, introduction to Mico Carboni, *Ettore Sottsass* (Orléans, 2006), p. 7.

34. Frank Gehry cited in "A Conversation with Gehry: The Design Process," interview by Beatriz Colomina, *El Croquis* 117 (2003), pp. xvi–xvii.

35. Achim Menges, "Material Computation: Higher Integration in Morphogenetic Design," *AD Architectural Design* 82, no. 2 (2012), p. 122.

36. Lovegrove 2004 (see note 19), p. 83.

37. Alisa Andrasek, "Open Synthesis: Toward a Resilient Fabric of Architecture," *LOG* 25 (Summer 2012), p. 47.

38. Ross Lovegrove, "Super-Natural: Parametricism in Product Design," in Lovegrove 2004 (see note 19), p. 107.

39. Lovegrove, in conversation with the author, December 2016.

40. See John Frazer, *An Evolutionary Architecture* (London, 1995).

41. Lovegrove 2004 (see note 38), p. 107.

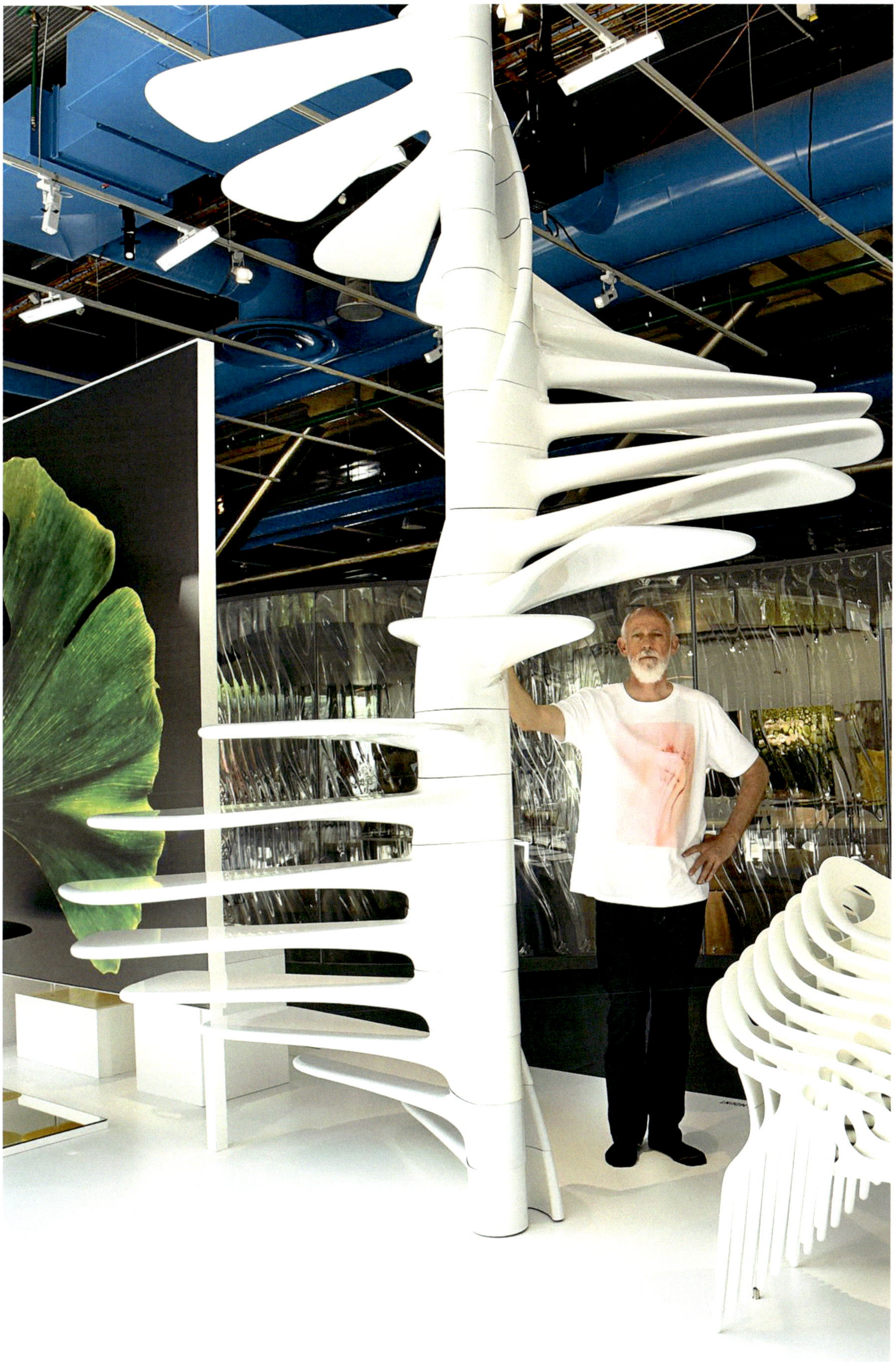

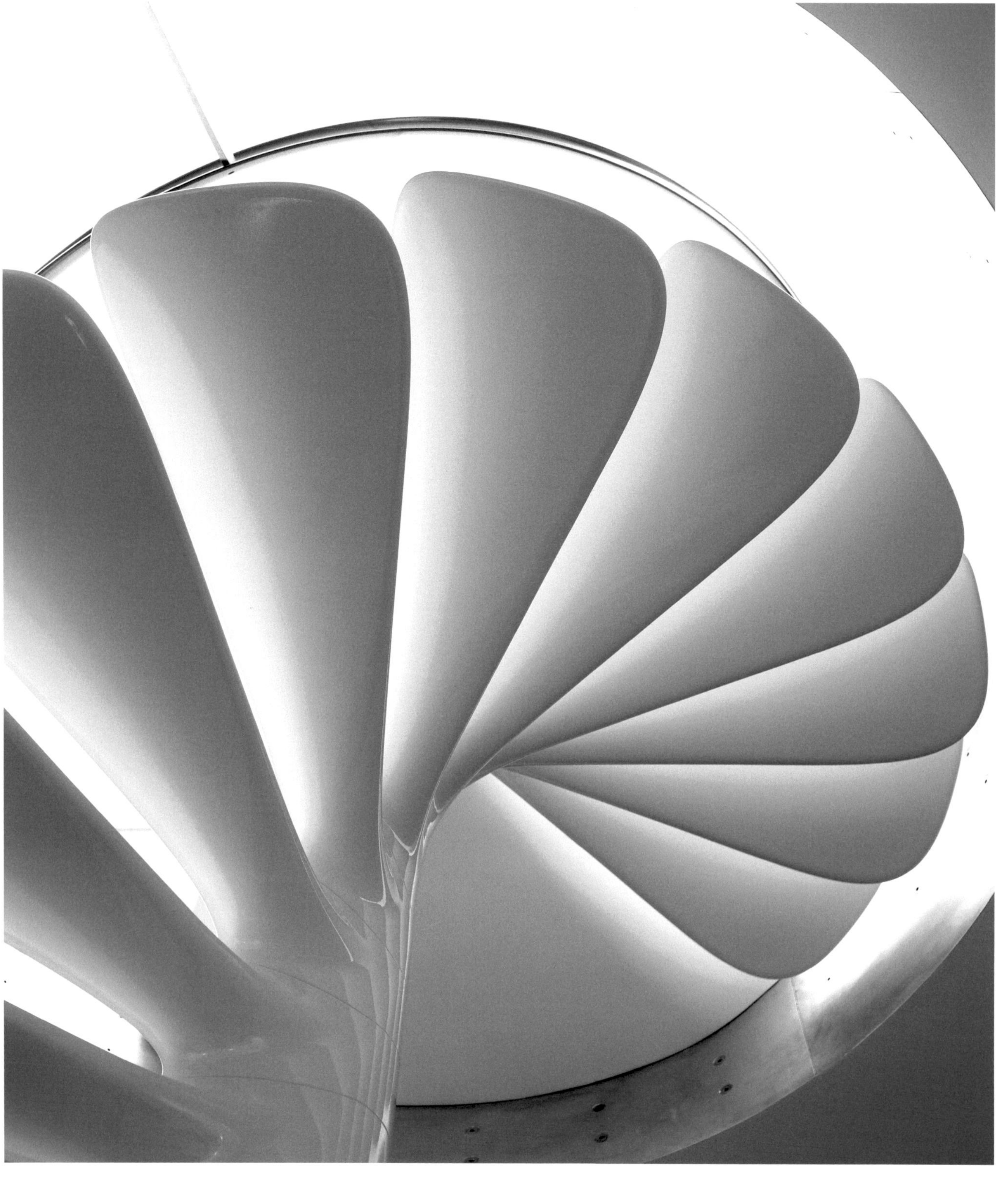

DNA staircase, 2005 (detail)

See pp. 56–57

Organic Essentialism

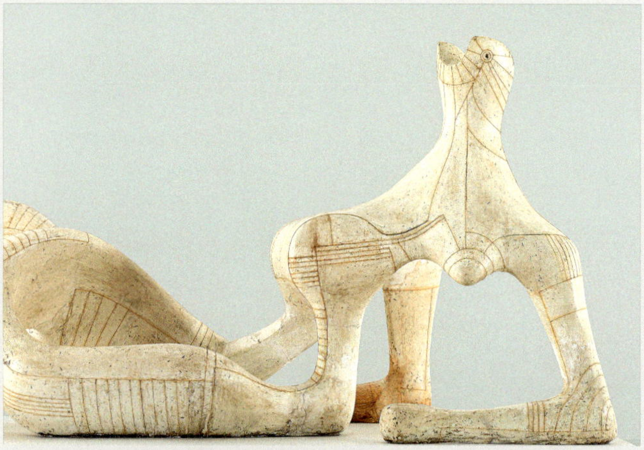
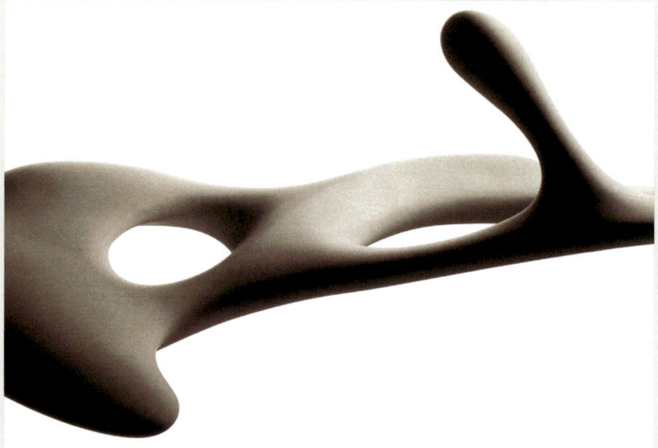
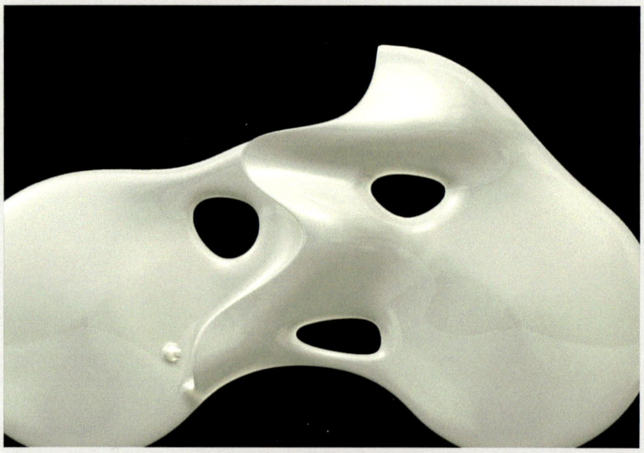
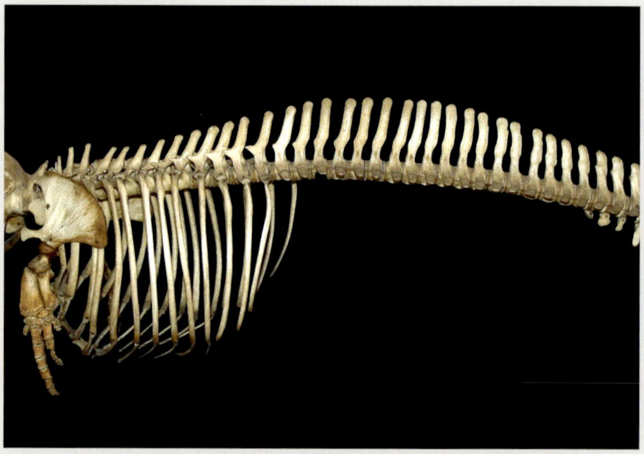
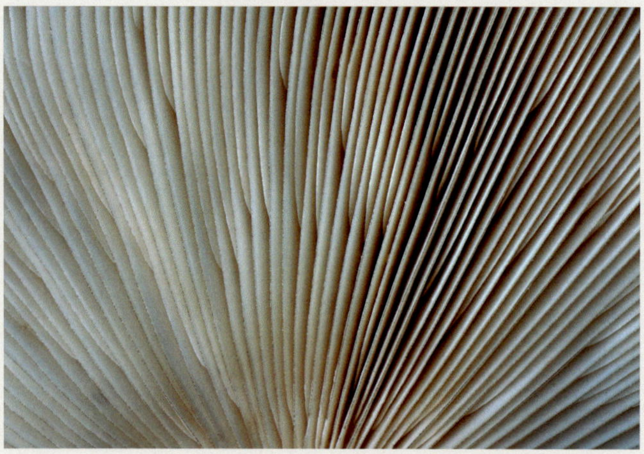
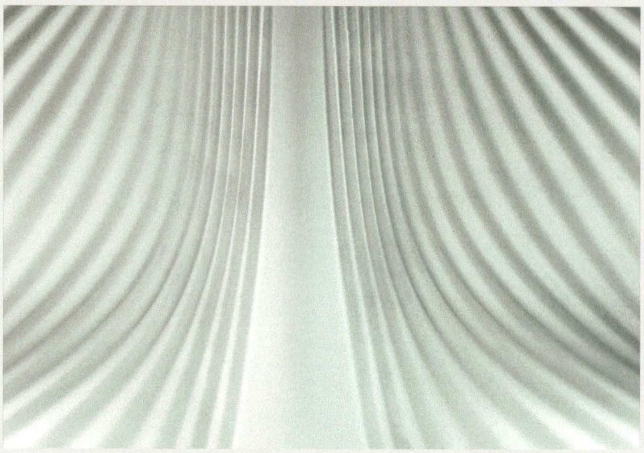

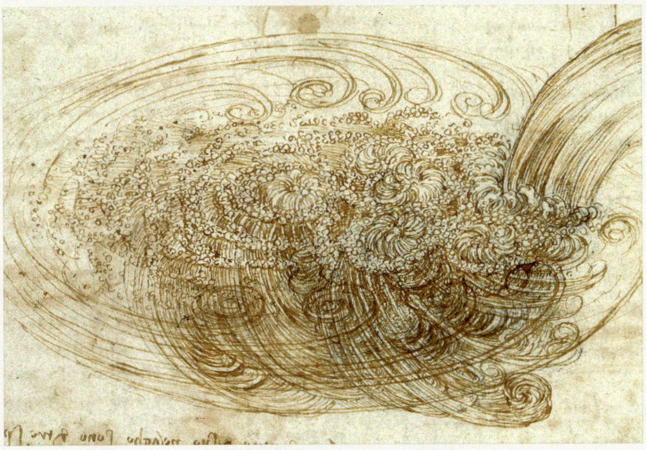

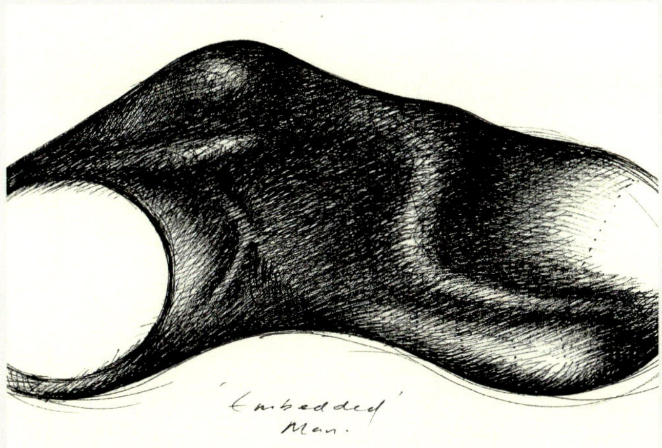
'Embedded'
Man.

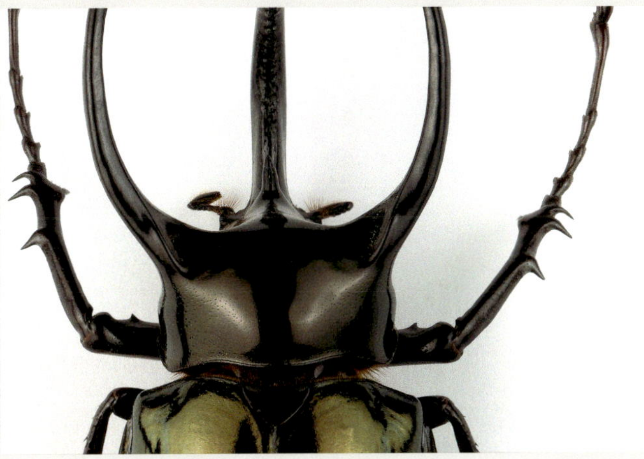

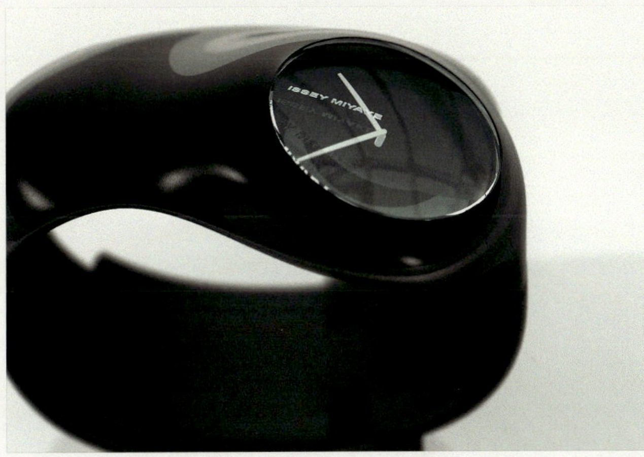
ISSEY MIYAKE

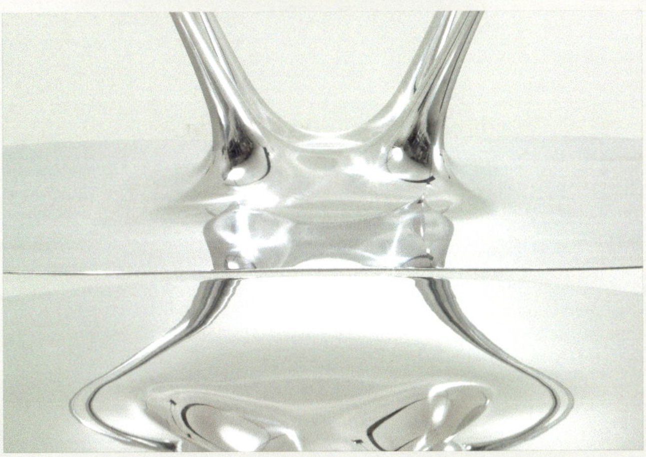

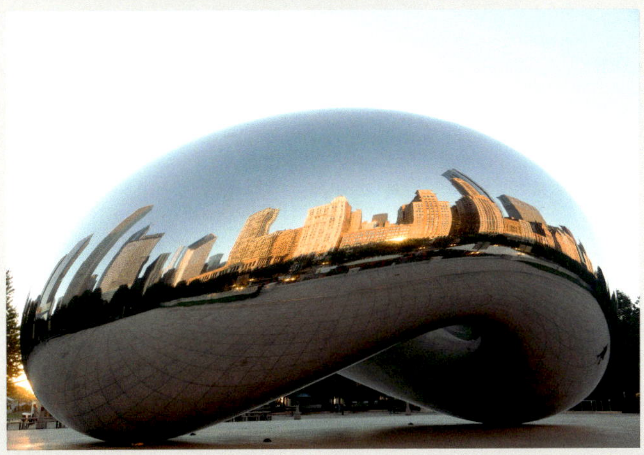

Unless indicated otherwise, all the works in this group have been produced by Ross Lovegrove. The lenders of works are mentioned in the "List of Exhibited Works," pp. 152–153.

Page 46 *(from left to right and top to bottom)*

Henry Moore, *Reclining Figure* (detail), 1951, plaster and string, 105.4 × 227.3 × 89.2 cm, Tate, London

Ross Lovegrove, *Biowood* (detail), 1999, wooden sculpture, 10 × 42 × 10 cm, Ross Lovegrove Collection

Ross Lovegrove, *Ridon* (detail), 2005, carbon fiber sculpture, 105.5 × 181.5 × 40 cm, Ross Lovegrove Collection

Skeleton of a large dolphin (*Tursiops truncates*), Zoological Museum at the University of Coimbra, Portugal

Gills of the Lepiote deguenillee (*Macrolepiota rhacodes)*, Leicestershire, United Kingdom

Ross Lovegrove, Biophilia chair (detail), 2014, rotomolded polyethylene, 103 × 47 × 56 cm, manufacturer: Vondom, Spain

Page 47 *(from left to right and top to bottom)*

Leonardo da Vinci, studies of flowing water, with notes, ca. 1510–1513, pen and ink over traces of black chalk, on paper, 29.6 × 20.7 cm, Royal Collection Trust

Ross Lovegrove, *Ridon* (detail), 2005, pen on paper, 105.5 × 181.5 × 40 cm, Ross Lovegrove Collection

Atlas Beetle (*Chalcosoma atlas*), can measure up to 12 cm, Philippines

Ross Lovegrove, Hu watch on its stand (inspired by the long benches in the Liquid series), 2005–2008, titanium, 10 × 5 × 10 cm, manufacturer: Issey Miyake Watch, Japan

Anish Kapoor, *Cloud Gate*, 2004, Millennium Park, Chicago, United States

Ross Lovegrove, Megabioform table (detail), 2000, aluminum, 72 × 290 × 162 cm, Ross Lovegrove Collection

Underpinning Ross Lovegrove's creative and experimental approach, the concept of "organic essentialism" is thought of as a "base code" that responds to different criteria in order to imagine new types of objects. With the resources employed for the conception of an object up until its use, optimization is found at every level. The use of innovative materials and the reduction of forms to their essential components originate in the search for efficiency, inspired by models from nature. Biomimetics, associated with algorithmic computer language, lead to complex structural combinations that allow one to obtain the desired form with an extreme precision. As early as 1994 Ross Lovegrove developed these principles through a suite of office furniture, the Emma Office System. In 1998, he applied this principal of material reduction (which he calls "fat-free") to the Go Chair, which allows the structure to reveal itself. In 2000, an in-depth study of the possible uses of copper allowed him to create the Cetacea lounge chair. Five years later, the DNA staircase combined the structural strength of bone structure and the double helix form of DNA to formulate new principles of architectural conception. The Ginko table (2003–2006) was based on the study of organic growth to give a solid surface the appearance of a plant membrane. Working with the properties of sound, the designer proposed, with the Muon speakers (2005–2007), forms that combine the dynamics of sound waves and the movement of water. Offering a sensory experience at the frontiers of synesthesia, the design demonstrates the "intelligent and synergetic interaction between the object and the effect" sought by the designer, as well as his way of interrogating the materiality of his works.

Olivier Zeitoun

Organic Essentialism

Ross Lovegrove

Organic essentialism is a base code that lends three-dimensional design succinct properties relating to reductionism and the elimination of extraneous mass in designed objects. It is a guiding principle in my work; it is not prescriptive but instead enables me to orientate and hold true from deviation the objective of creating leanness and economy in organic form. It also relates to the trinity of materials—material, structure, and function—often selected for their holistic containment rather than piece-part construction. This method of construction is quickly becoming a redundant vestige of an industrial age, existing in an accelerated state of transition from the mechanical toward the biological.

If we talk about form, there is an implication that a silent, static object or product can be infused with an energy that is in a state of subtlety perceptible, generative motion.

Forms created in nature from natural forces emerge from intrinsic forces; this fulfillment is then arrested by extrinsic forces. The resulting forms and structures are in a symbiotic relationship with elemental, atmospheric forces such that they exist in a pure harmonic equilibrium between material mass, purpose, and scale. This generative approach is thus an invisible force that can help an inanimate, manmade object be immersed in a natural world.

I have seen and felt this in the work of Henry Moore or Anthony Gormley, whereby the work of art is so present that it becomes proprioceptive with its own awareness of space and time and the impact it has upon them. This quality is extremely rare in industrially made artifacts, which, by definition, are not enduring—due to their inbuilt obsolescence—and therefore appear disposable and ephemeral.

Organic essentialism is where instinct and awareness meet with a certain order that allows them to be defined; even if it appears expressionistic, it does not appear overly so, but only as if it already existed and a fine mist were sprayed into space to reveal it.

By definition, design must begin with a degree of arbitrary abstraction in order to initiate the process of discovery that goes beyond existing references and then allows instinct combined with knowledge to fuse and solidify, in accordance with what I believe is an evolutionary gene of creative consciousness. This method of revealing the inner truth of form is directly related to the harmonic accord of evolutionary morphology. Understanding is not necessarily knowledge.

Rather, it is a response that leads to an inversion of one's sense of being, turning into oneself to find honest, unpretentious "inner feelings," which are not about seeking commercial design with premeditated signs of attraction.

Attraction or attractiveness is a byproduct of creation that possesses harmonics, economy, and a sensuality born of the universal, deep primordial, polysensorial memory that some of us have retained. Indeed, if we trace the accent of form, beginning with insight into the way that the organic landscape and materials of our early habits first penetrated man's imagination, we can begin to accept and understand the profound effect organic form has had and can have on our consciousness and sensibility. Organic form remains at our collective core as a reflexive, prelinguistic understanding, connecting our organic being with the plasticity of art, design, and emergent architecture. There is no objective or agenda associated with organically essential objects because they are in fact just paused moments of accelerated evolution, converged and made physical by the creative mind.

My designs are never so dematerialized that they lose their physical presence, but the research I undertake with my sculptural works or limited editions, such as my *Bioform* series or carbon *Vertebrae,* is undertaken to explore the limits of this territory using materials and technologies that facilitate rare scale, emotional impact, or the abstraction of function. For example, there is a conscious agenda for formal purity, as in the work of Richard Serra or Anish Kapoor, because this suggests an aesthetic summit within the concept I am exploring as a form of luxury that distances itself from the common man.

It is important to understand that owning a work of art to which access is limited due to its uniqueness and singularity, which also define its value, is a human need that, in fact, is diametrically opposed to the high-volume, high-commercial objectives of industry. Owing to endless mass production, industry often reduces geometry to its lowest common denominator, save the automobile or aerospace industries, which invest much more in intelligent, biomorphic, holistic integration.

It is important to look back to the very beginnings of human existence, when tools were fashioned by homo faber from a limited spectrum of organic materials to arrive at incredibly sincere and elemental objects that supported life itself. As the first humans became dependent on their success, they entered into a continuous endeavor of making and improving, edging toward art and complexity once fundamental needs had been met. Certain conditions and pressures allowed such tools to remain in a mode of perpetual adaptation and evolution, albeit incrementally. What is surprising to learn, on a more abstract level, is that via some form of global consciousness, the same tools were formed from an innate universal order of creation across the inhabited continents, as we see from prehistorical and archeological artifacts the world over. As we face the need to reduce and rationalize the amount of materials and resources we use today, the necessity of linking and correlating with these first tools is profound.

When I see such artifacts in a museum or hold them in my hand, they seem to transmit a form of energy, and it is this radiance that I would like to transpose into contemporary objects of industrial manufacture, as a way of communicating feelings about human transfer/continuum and exchange rather than a sense of disposability and transience.

My interest in evolutionary form as seen in species morphology and adaptation, aligned with contemporary advances in computational design and additive manufacturing, derives from my belief that this sort of succinctness and the natural trinity between structure, materials, and function—so omnipresent in the natural world—represent the key to the convergence between the natural and manmade environments with respect to integrated sustainability.

The works I show in the present exhibition and refer to in this catalogue are initial suggestions for developing a deeper thesis of investigation, supported by computational design to define a way of using organic essentialism as a foundation for new universal principles in creating tools for living. Communicating beauty and logic in parity, such tools satisfy scientific and artistic minds alike—from bicycles and prosthetics to driverless cars and the unforeseen devices of our collective future, which will no doubt be extensions of our mental space and physiognomy.

London, December 30, 2016

Cetacea lounge chair

2000
Hand-formed copper: 63.5 × 182.2 × 70.5 cm
Edition of 3 and 2 artist's proofs

Cetacea is a lounge chair whose design follows the principles of organic essentialism and is distinguished by having been realized in copper. Ross Lovegrove devotes particular attention to this material; his discovery of it has played a major role in the anthropological aspect of his work and marked certain crucial stages in the history of his technique. The name of the design was proposed by an antique dealer specializing in the sale of bones. Upon encountering the object for the first time, he instinctively recognized that it bore similarities to the large aquatic mammal family, and more specifically to the shape of a whale's jawbone. For Lovegrove, Cetacea is the symbol of the convergence between the physical properties of nature, which can be so spectacular and unexpected, and the effort produced by man to better understand it, and therefore to accord it greater respect.

MV

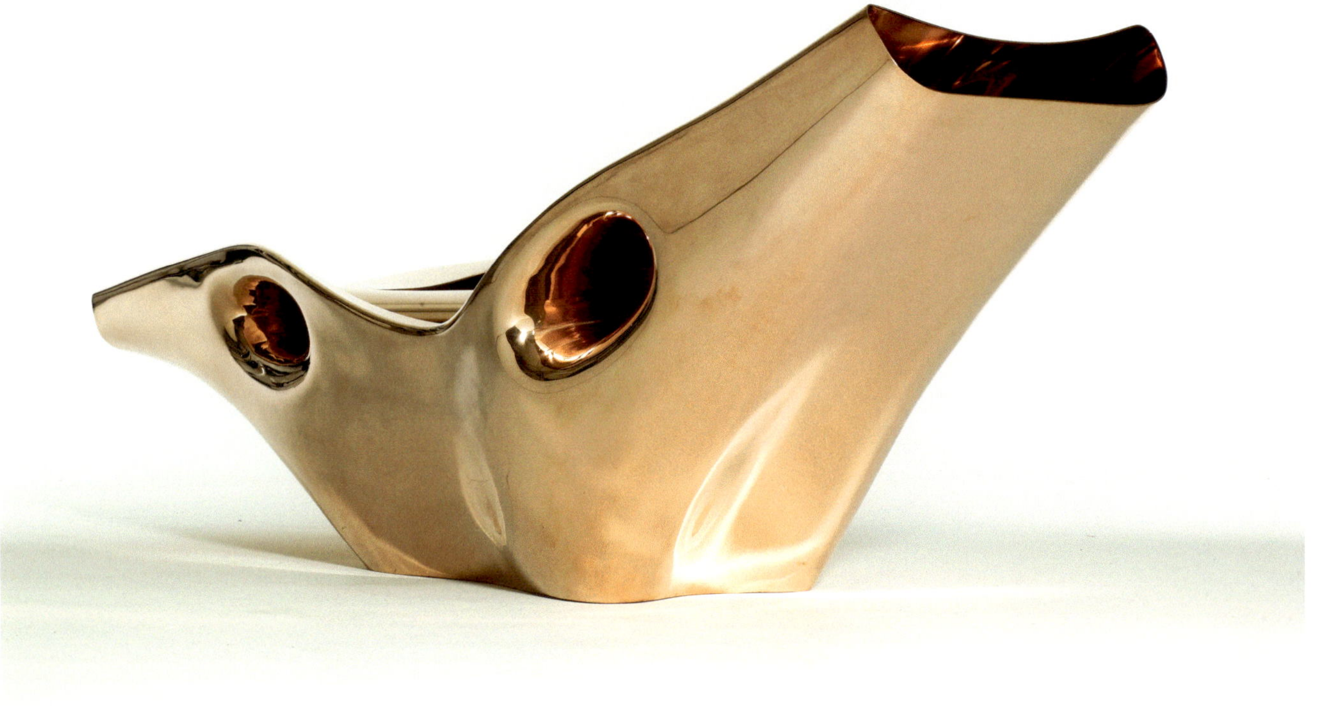

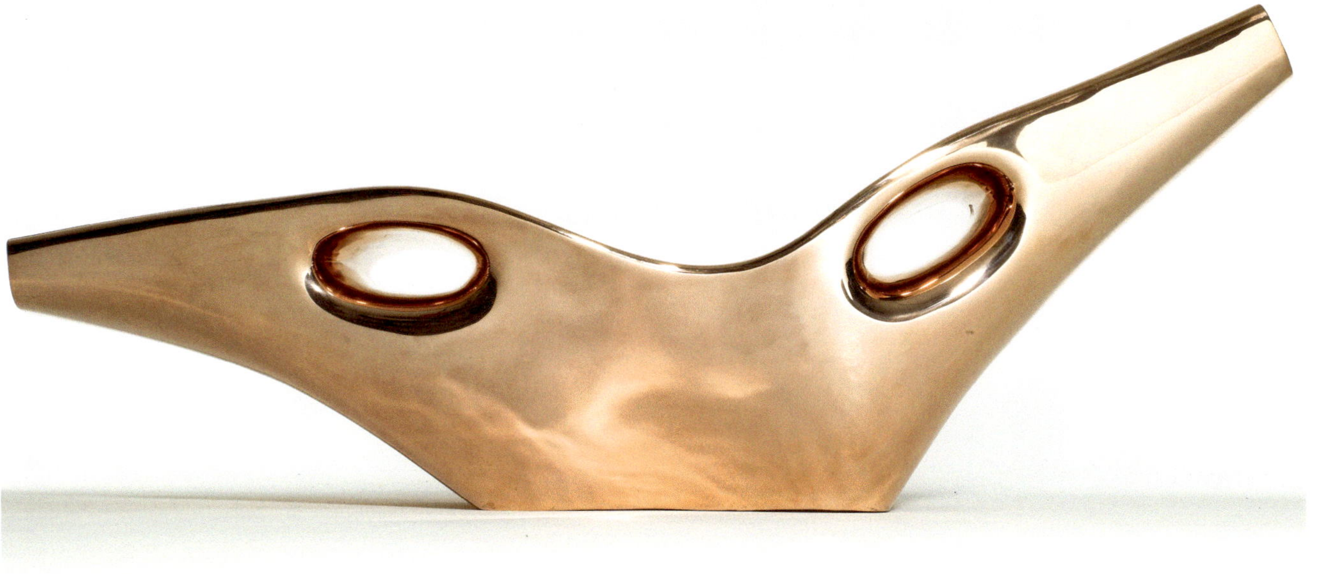

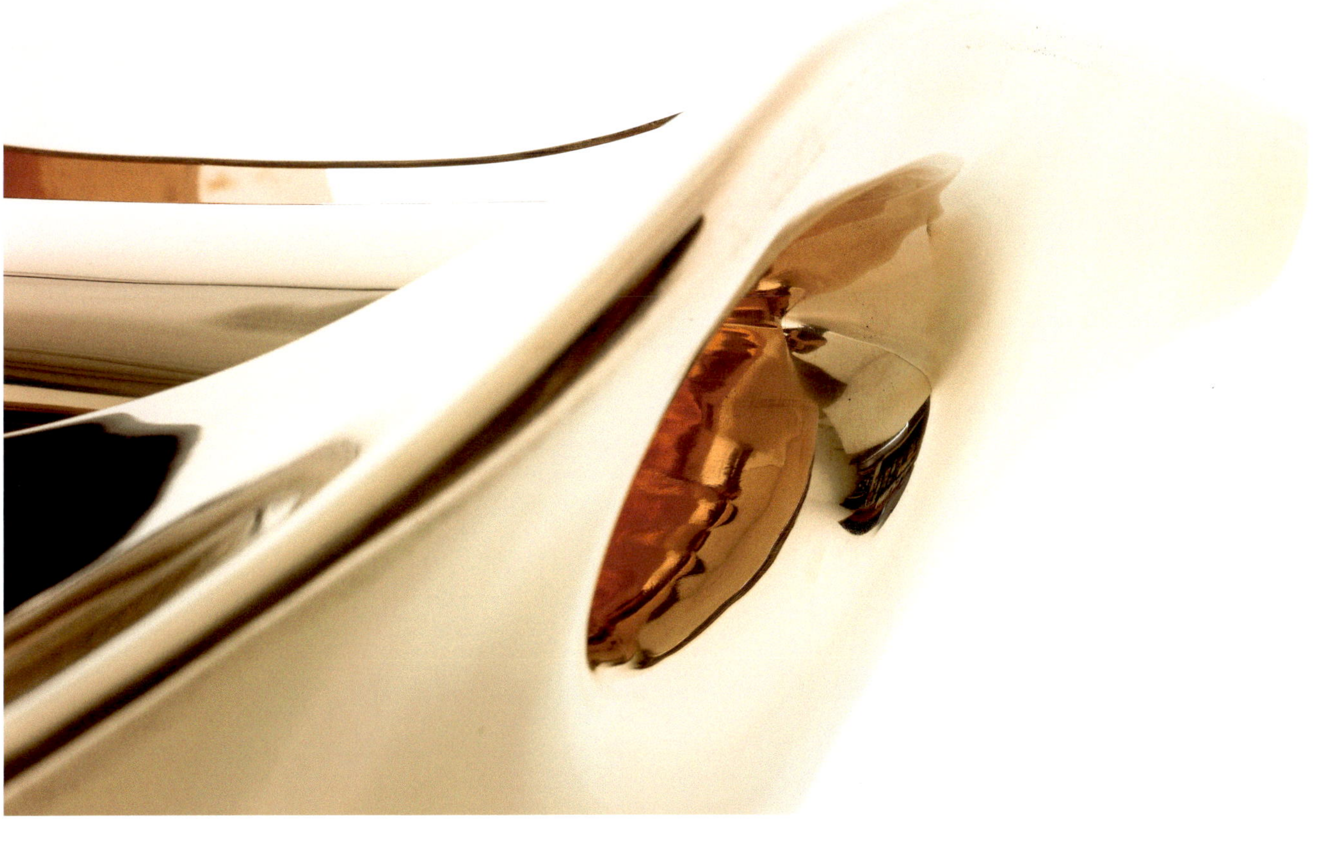

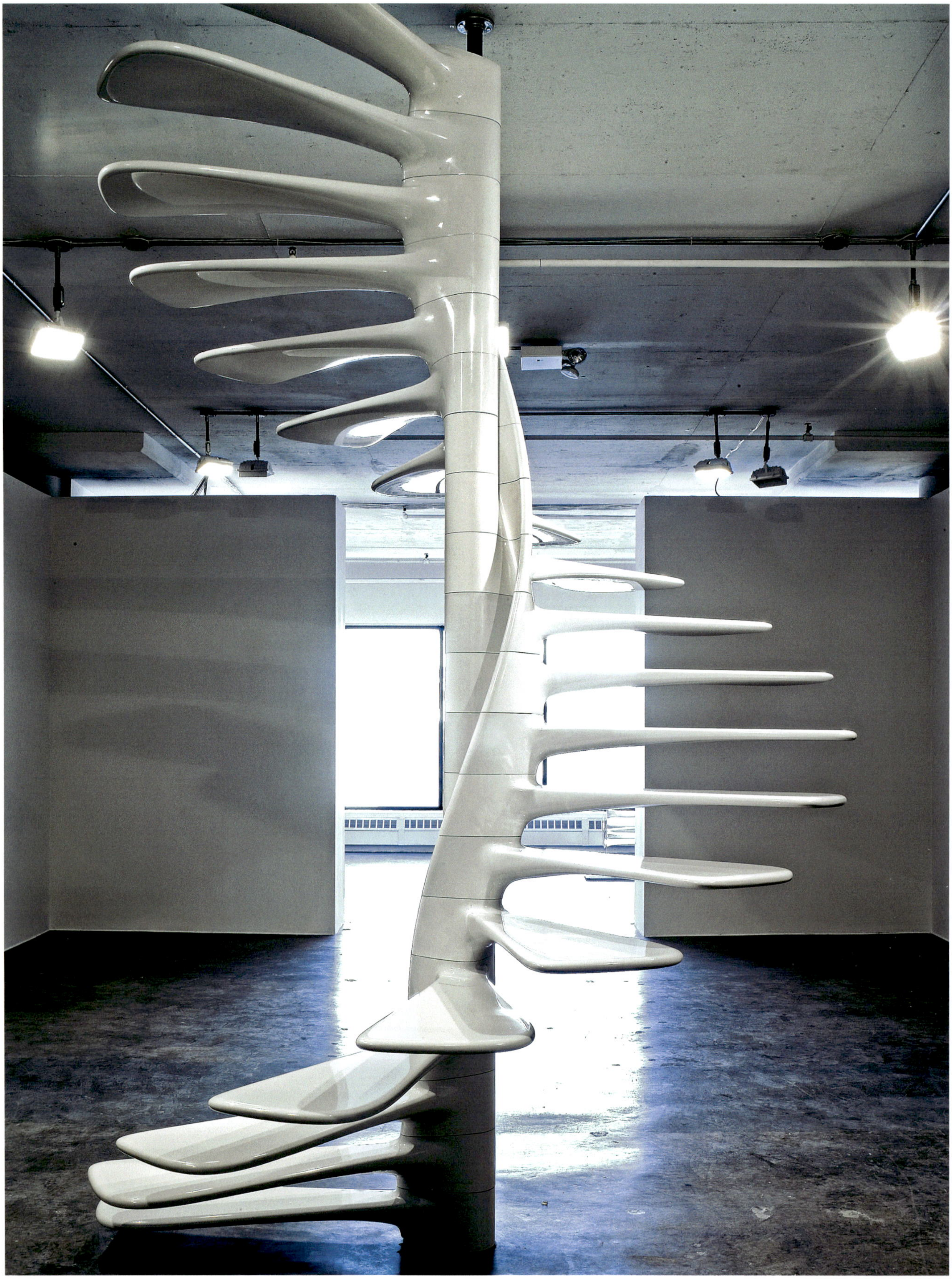

DNA staircase

2005
Combination of fiberglass and unidirectional carbon
H.: 371 cm, Dia.: 231 cm
Edition of 1 and 1 artist's proof

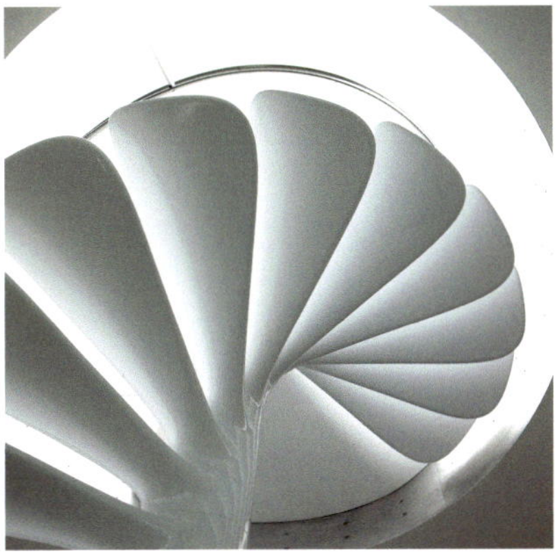

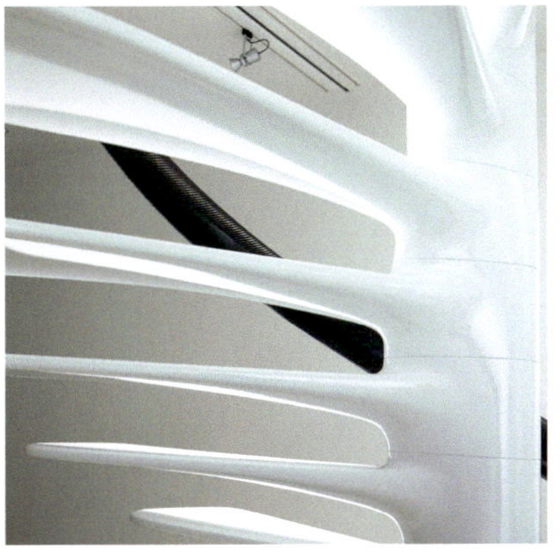

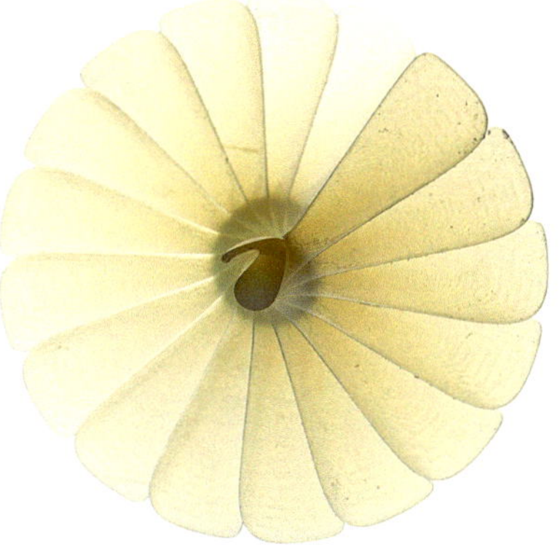

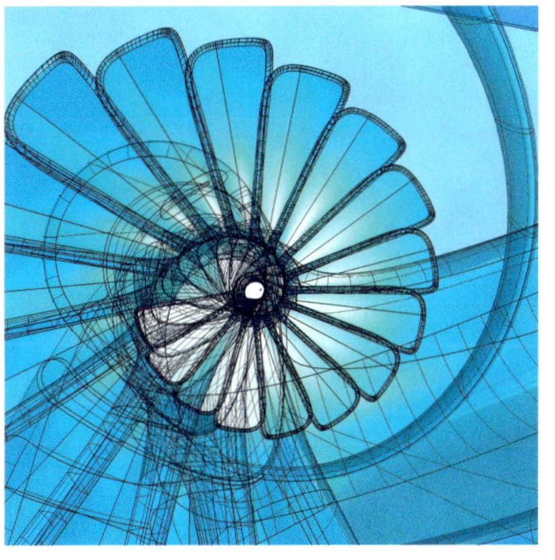

Bringing together the double helix of DNA and a bone structure, the DNA staircase allows free-flowing circulation over several floors at the heart of Ross Lovegrove's studio in London's Notting Hill district. It was made with compressed molded fiberglass and carbon fiber, a blend of materials that combines strength with flexibility, lending the design considerable structural force. The "essential" form of the staircase and its obviously organic nature reflect the designer's "organic essentialist" principles, which unite engineering and biological evolution and apply them to new principles of architectural design.
MV

Go Chair

1998–2001
Magnesium molded by injection,
magnesium powder coating, polypropylene seat
77.5 × 58.4 × 68.6 cm
Manufacturer: Bernhardt Design (United States)

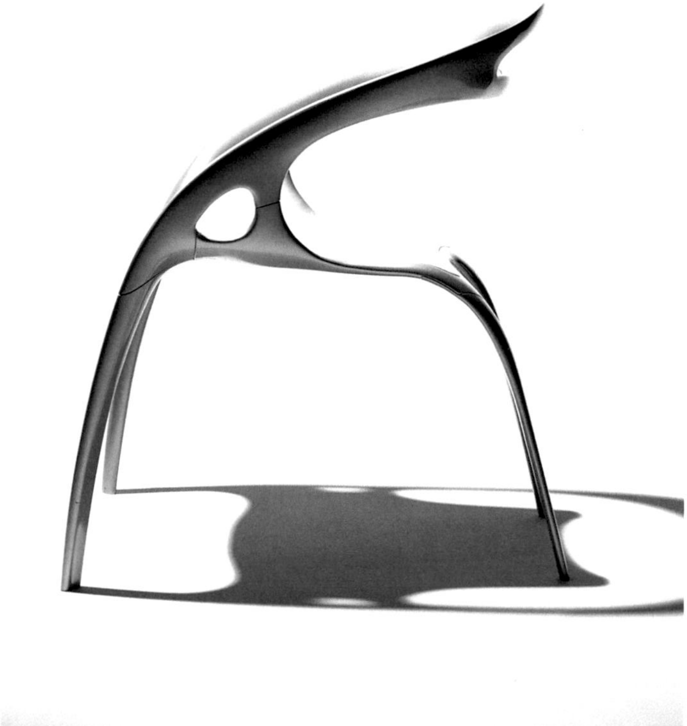

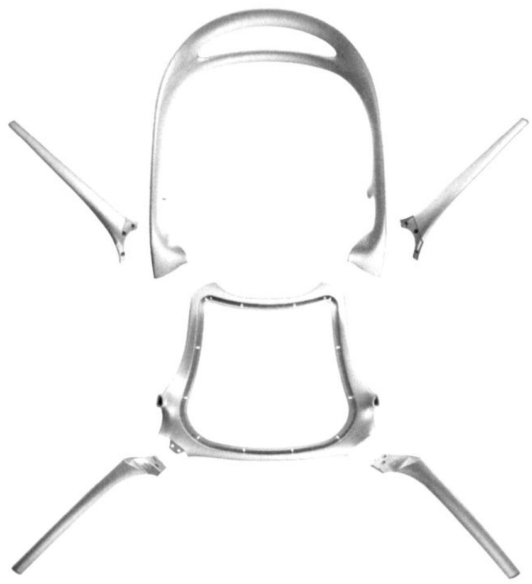

The material of the Go Chair makes it distinctive: it is almost entirely made of magnesium molded by injection. Only the seat is made of polypropylene or varnished wood, which makes it at the same time stable, light, and solid: the frame is covered in a layer of silver or white powdered paint. The first piece of furniture made using this technique, the Go Chair encapsulates the notion of "organic essentialism" implemented by Ross Lovegrove: the forms are not constructed, but rather "interlace" following a natural process. "Fat free" according to the designer's terminology, reduced to its lines of strength, the "chair" evolves into a skeleton, alluding to the bone structure of an organism.
MV

New Nature floor lamp

2012
Steel, polymer;
H.: 188 cm, Dia.: 28 cm (base)
3-D plan of the project
Manufacturer: Artemide (Italy)

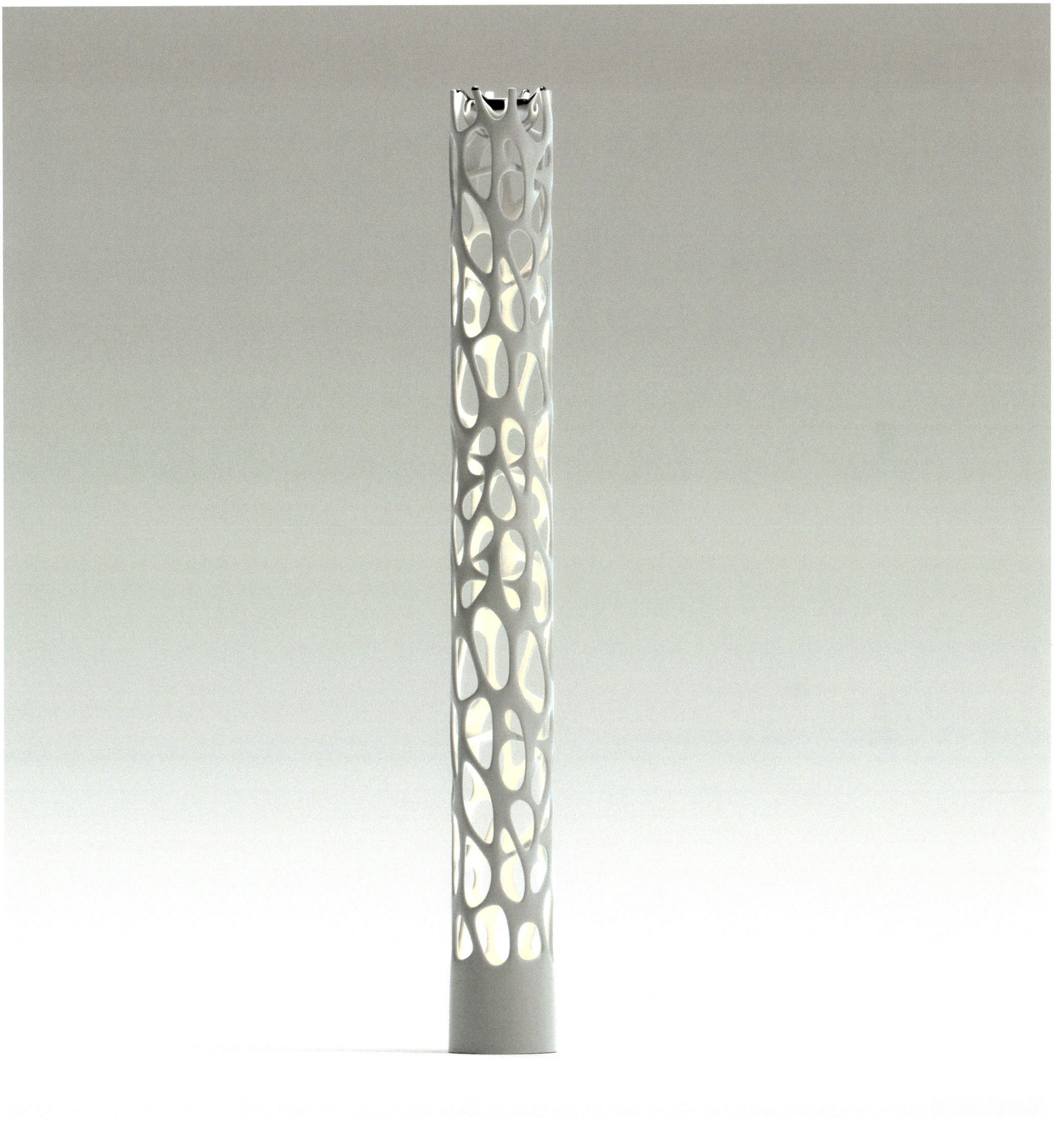

Supernatural chair

2005–2008
Single shell in polypropylene
79 × 64 × 56 cm
Manufacturer: Moroso (Italy)

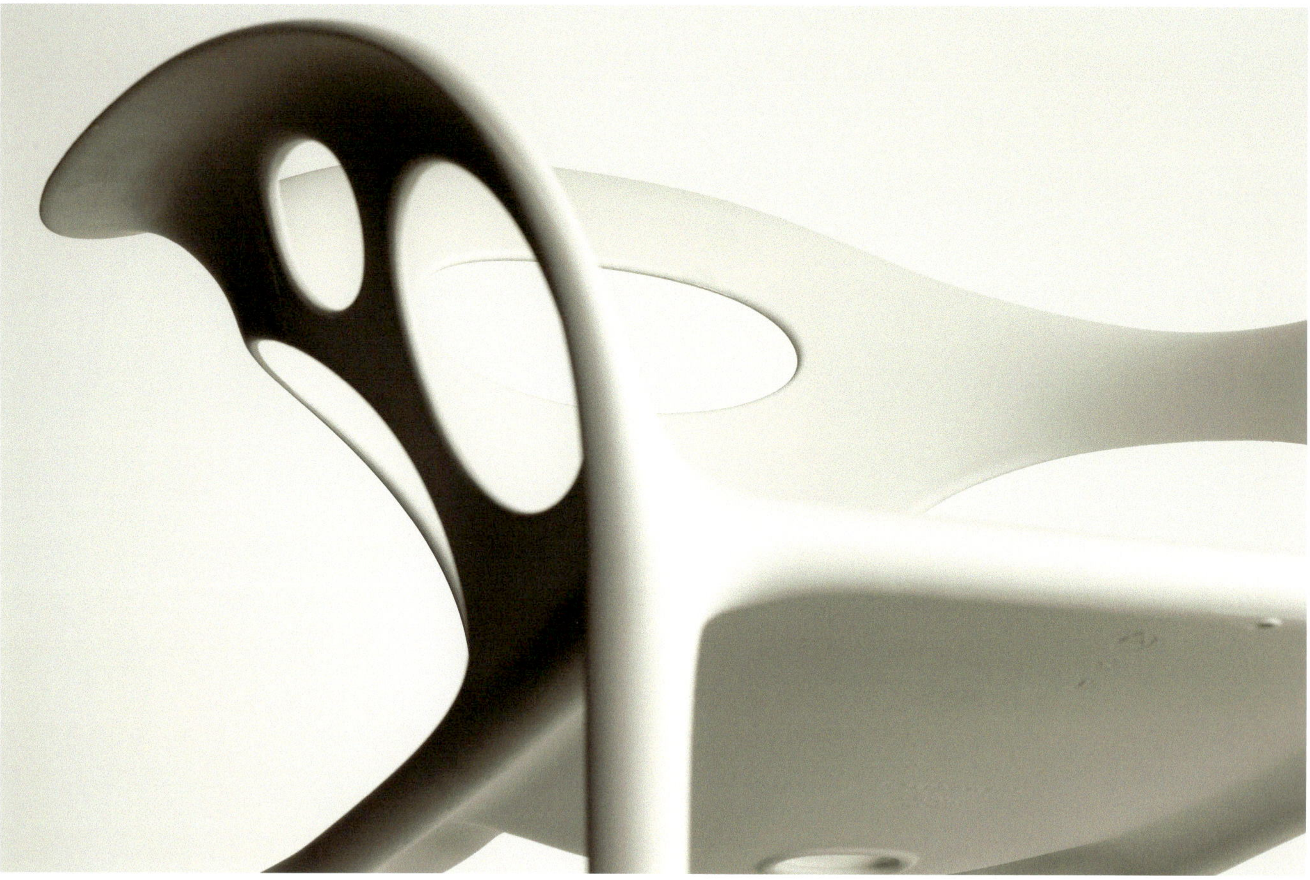

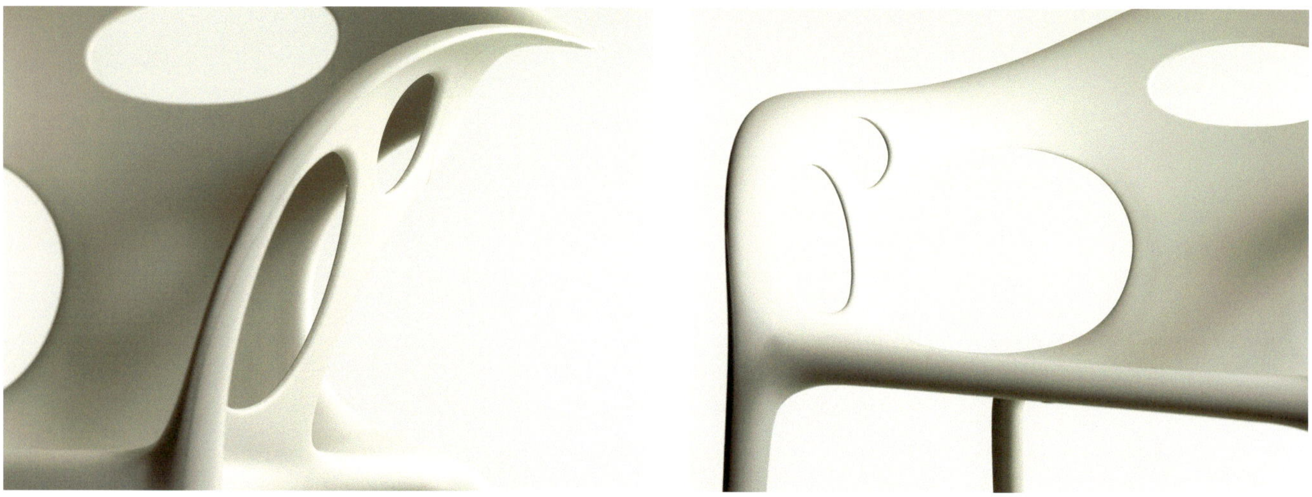

Andromeda light fixture

2006–2008
Suspended ceiling lamp
Steel with painted finish, hard polyurethane foam, mirrors, LED: 27 × 80 × 46 cm
Manufacturer: Yamagiwa (Japan)

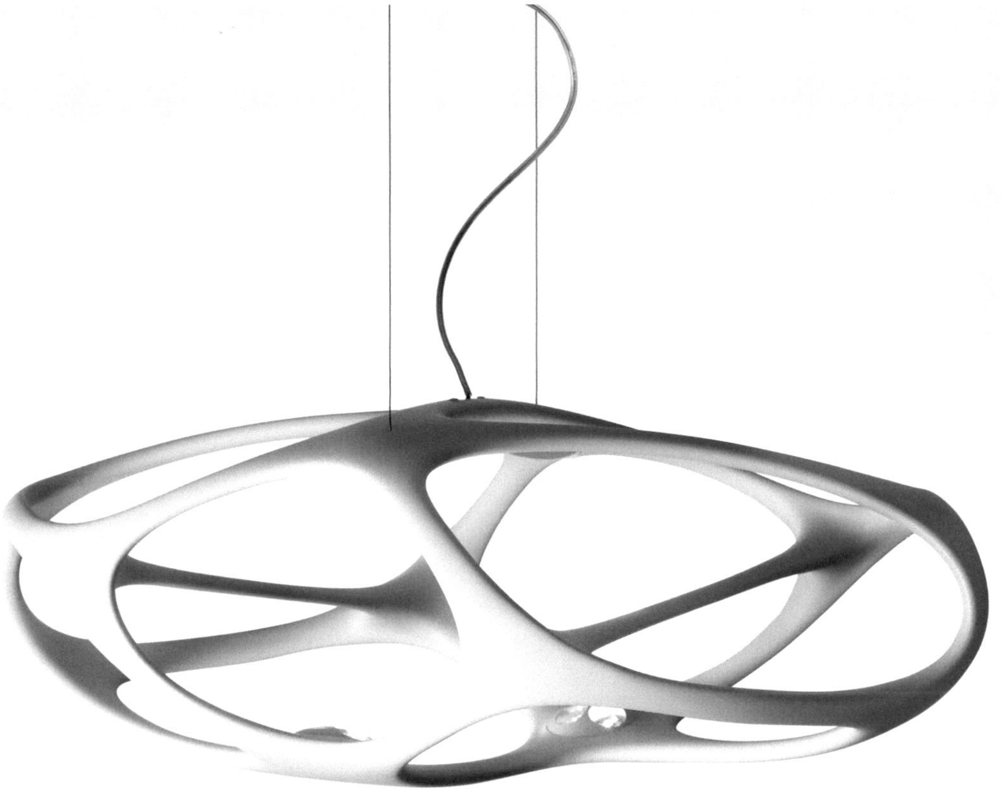

Developed for the Japanese company Yamagiwa, the Andromeda light fixture is made up of a bundle of LED lights embedded in a structure made of polyurethane foam with mirrors fixed by magnets. The light produced by the LEDs is reflected in the mirrors set at an angle to project moving shadows into space. The complex organic form of the Andromeda light is the result of sophisticated digital manufacturing technologies, which have allowed the maximum reduction of the physical mass of the object.

MV

Gingko table

2003–2006
Carbon fiber; H.: 72 cm, Dia.: 126 cm
Edition of 12 and 2 artist's proofs

The Gingko table, like the Liquid Carbon bench or the DNA staircase, shown during the *Endurance* exhibition (New York, 2007), is the result of the exploration of the structural potentials of carbon fiber and the generative processes of digital manufacturing technology. Graphite is a form of carbon that, combined with epoxy resins and other heat fixing elements, looks like skin. By layering several of these carbon "skins," each bearing different characteristics, an extremely resistant composite material is obtained, light and clear, which appeals to all the senses.
This table embodies all the principles that Ross Lovegrove holds dear, particularly that of "single surface deformation": the idea is that a simple surface can suddenly come to life and transform itself by following natural laws of growth. This "living" form could be realized thanks to the techniques used in the automobile and aeronautical industries.
Produced in a limited edition, the Gingko table, whose name harks back to *Ginkgo biloba*—the most ancient species of tree and the first to have grown back after the Hiroshima catastrophe—suggests the growth of a plant: it spreads the three petals of its surface around an invisible internal core, the central fold of the object being placed at the point where the forces of the material converge.
MV

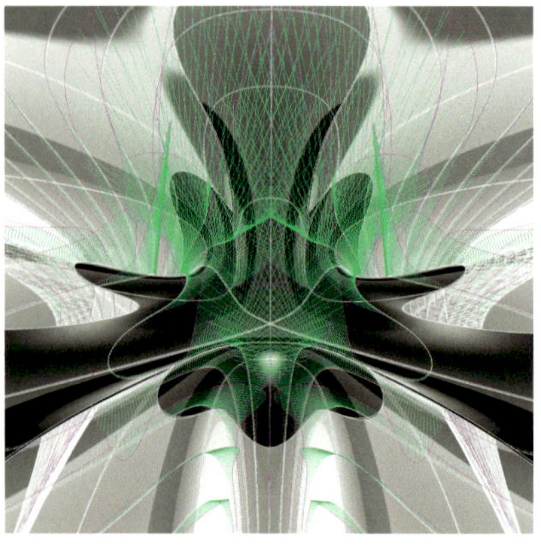

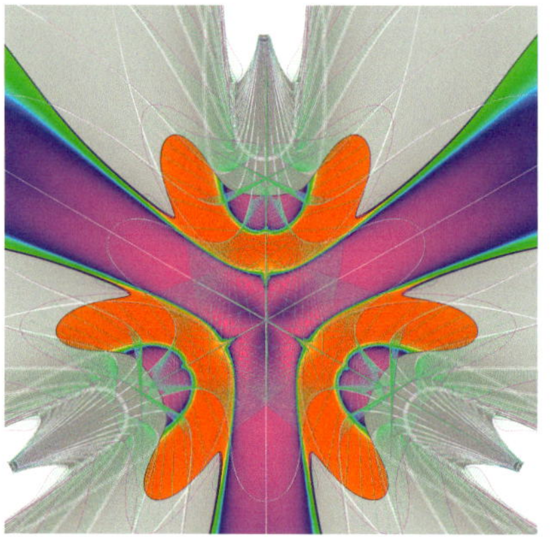

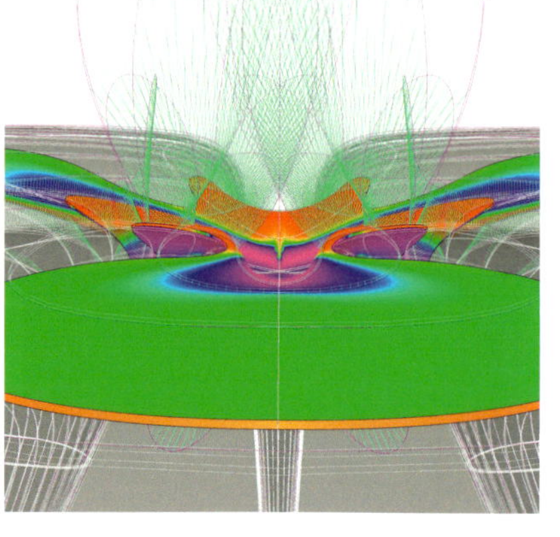

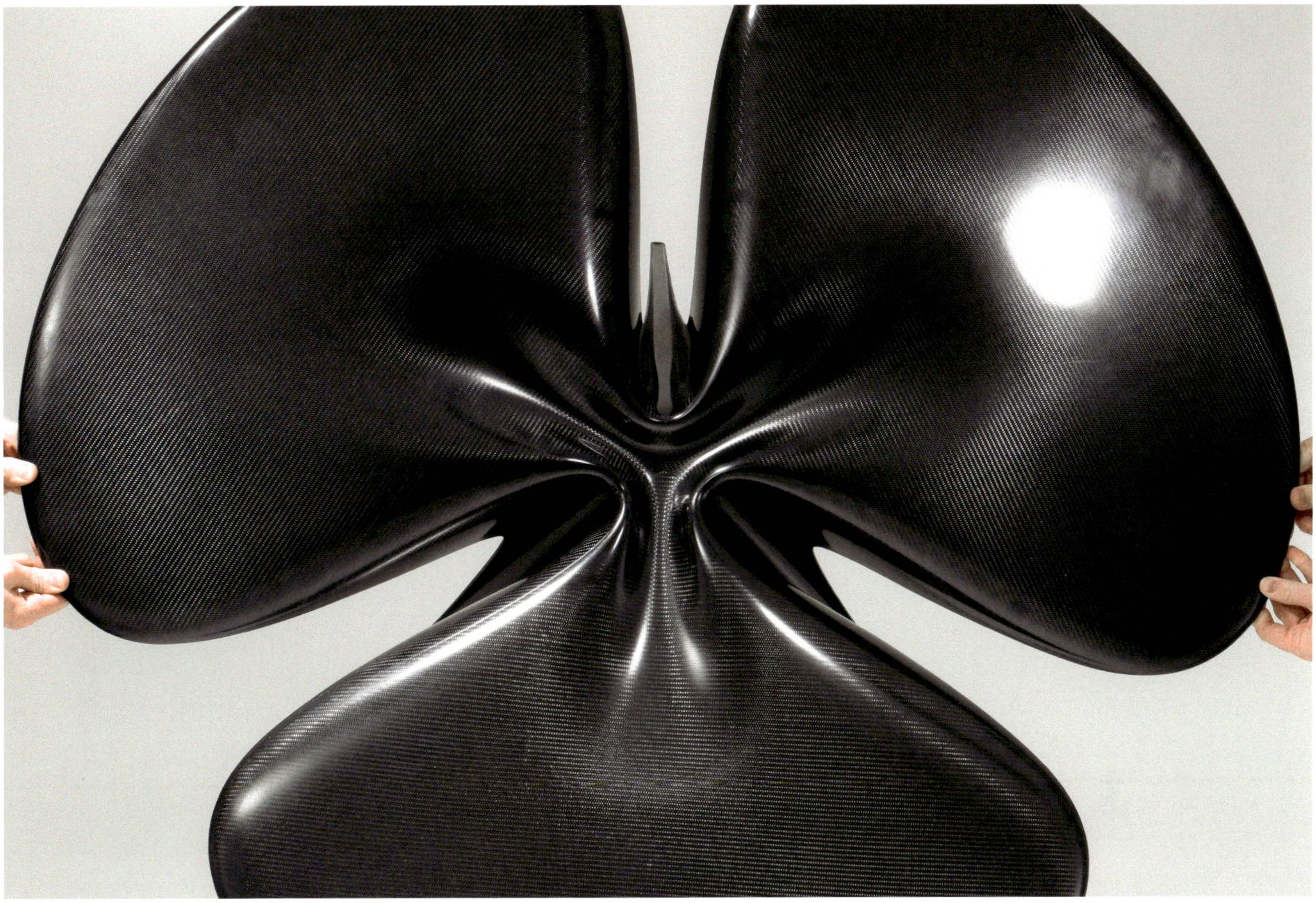

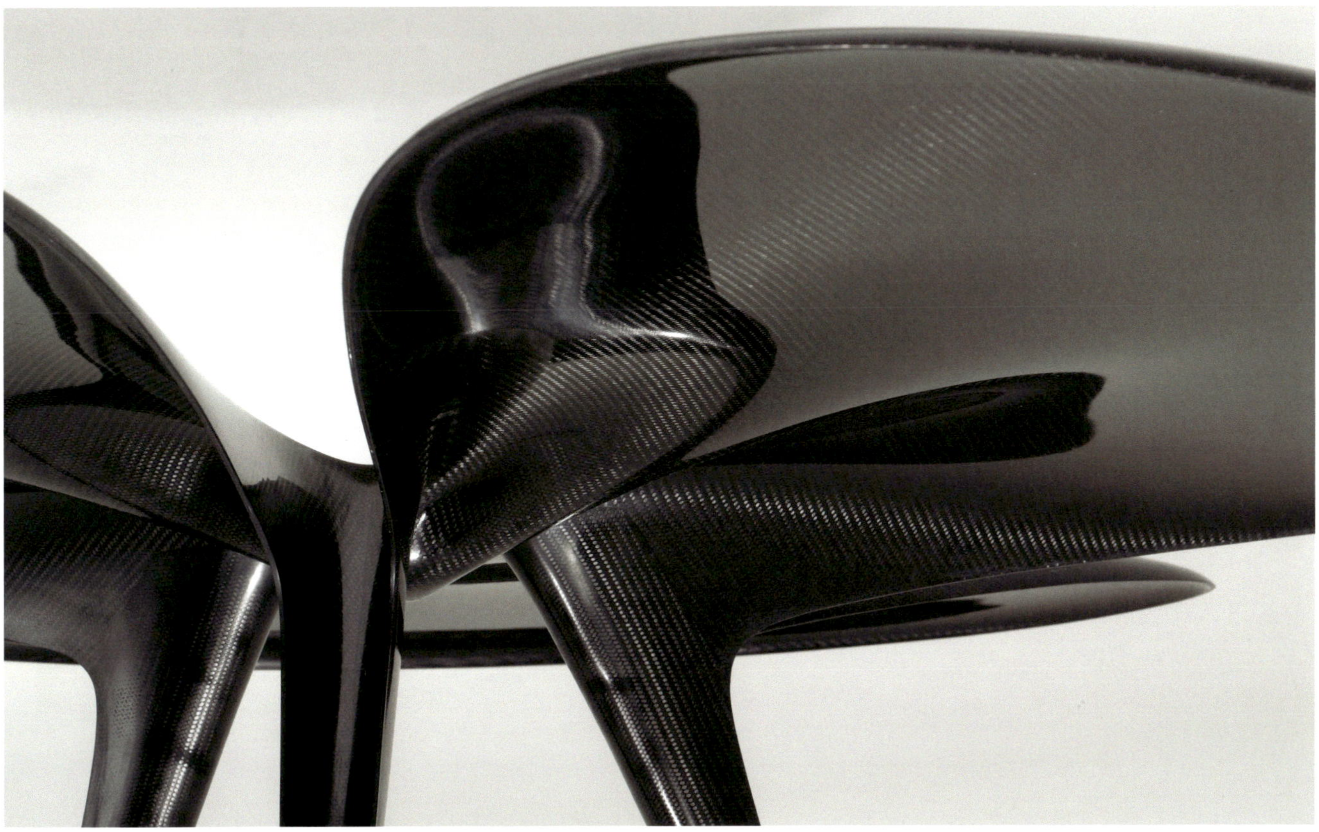

Ridon

2005
Carbon fiber on a digitally milled foam rubber core;
105.5 × 181.5 × 40 cm
Edition of 6 and 2 artist's proofs

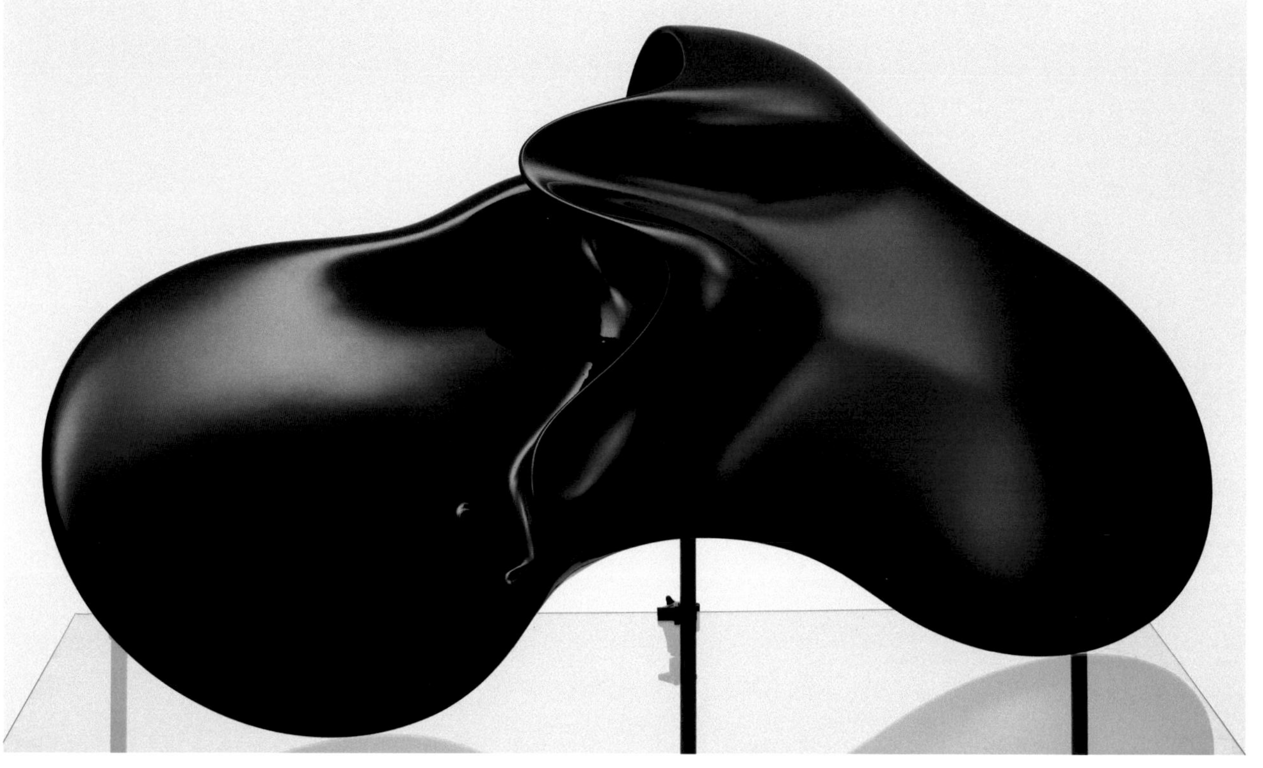

Ridon is a sculpture in carbon fiber on a foam rubber core, digitally milled, and produced in a total edition of eight. This extremely light and resistant material has enabled the production of a solidly compressed form of motorcycle, suggesting the fusion of man and machine at full speed. The piece recalls the dynamic that characterizes Umberto Boccioni's futurist sculpture *Forme uniche nella continuità dello spazio* (*Unique Forms of Continuity in Space,* 1913). Nevertheless, this fixed dynamic combined with the concept of speed resonates with Ross Lovegrove like a sense of doom that weighs on the future of man, in a society that encourages mobility and aspires to speed but increasingly restricts the conditions for its deployment. *Ridon* embodies the holy trinity close to the designer's heart: organic beauty, material reduction, and attention to ergonomics.

MV

Biolove bicycle

2003–2011
Carbon fiber, metal;
108 × 185 × 71 cm
Manufacturer: Biomega (Denmark)

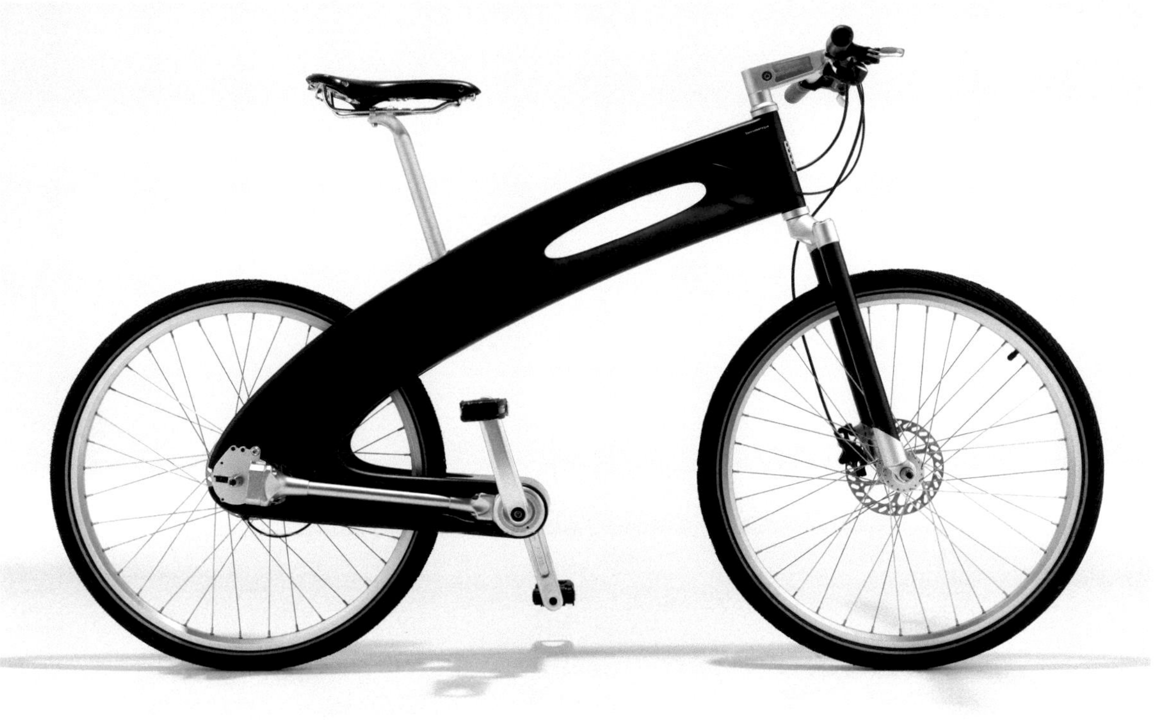

GT110 suitcase

2006–2007
Carbon fiber, Kevlar;
65 × 80 × 28 cm
Manufacturer: Globe-Trotter
(United Kingdom)

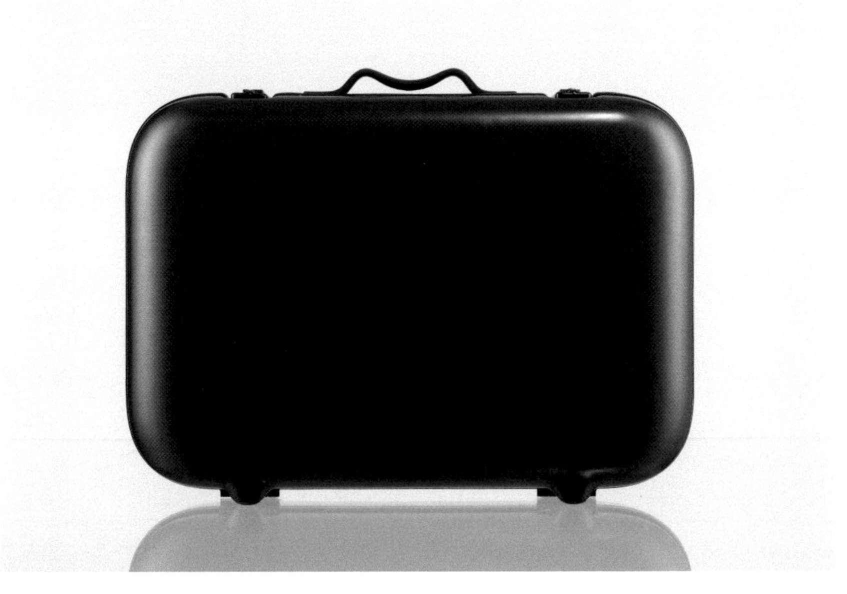

Moot chair

2012–2013
Carbon fiber; 80 × 59 × 46 cm
Edition of 58 and 2 artist's proofs
Manufacturer: Established & Sons
(United Kingdom)

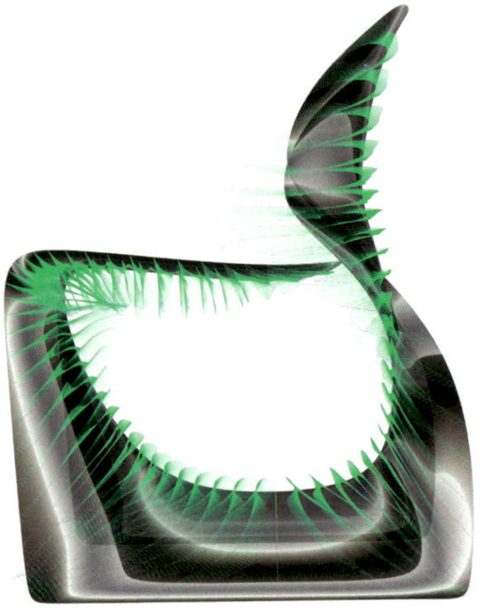

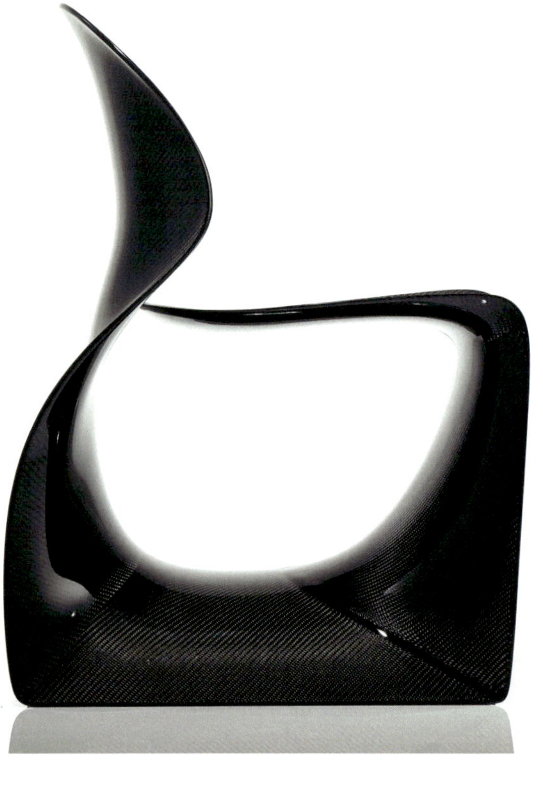

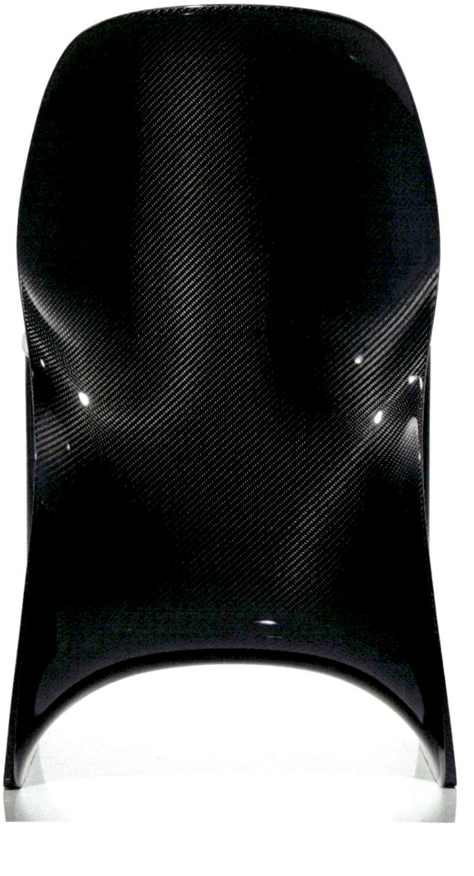

Liquid Carbon bench

2007
Carbon fiber on a digitally milled
polyurethane foam core; 60 × 285 × 85 cm
Edition of 10 and 2 artist's proofs
In the background: Megabioform table (see page 73)

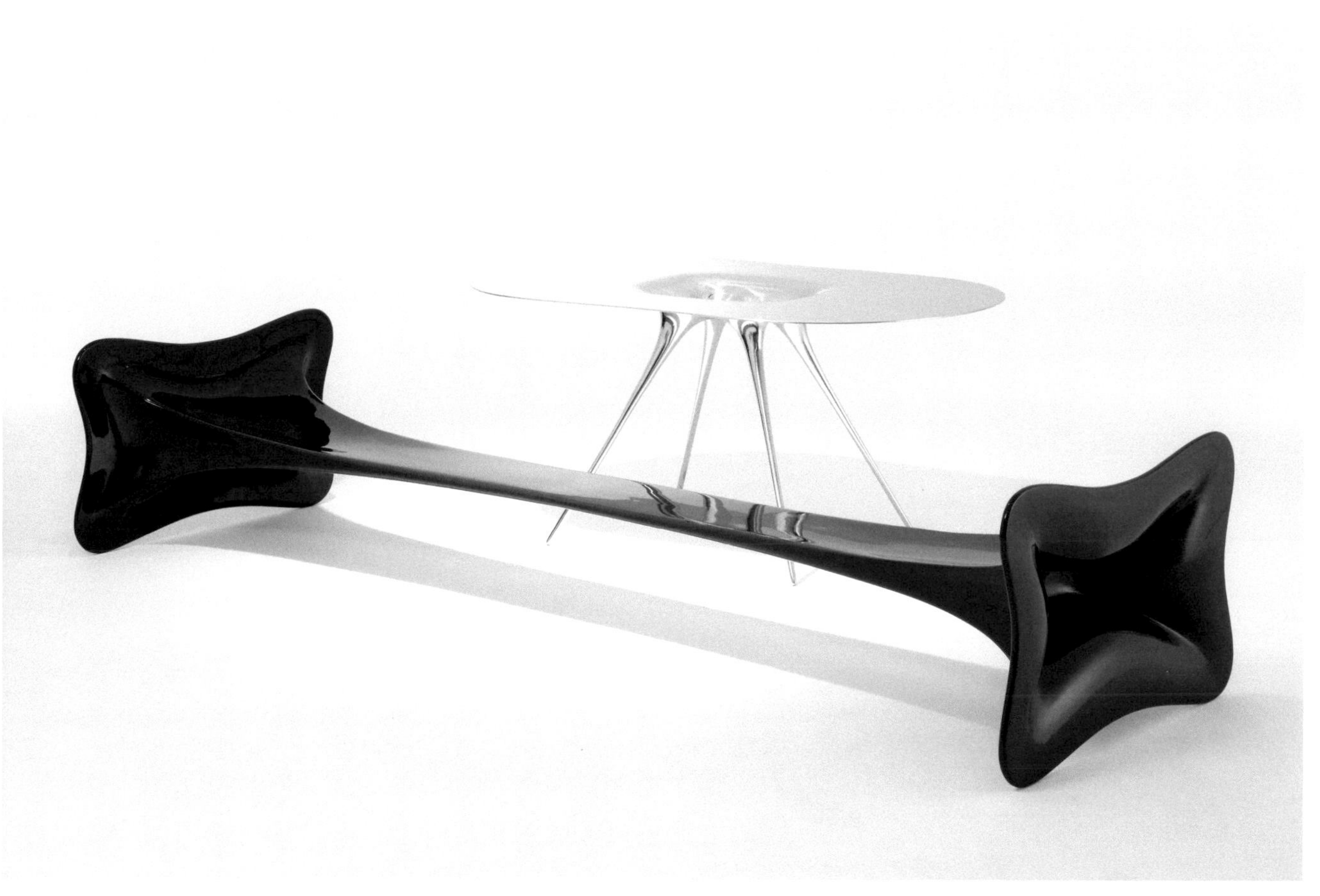

This bench made with carbon fiber and polyurethane foam creates the effect of a fluid and elastic form. This impression is produced by the animation procedure used in the making of the object. The technique allows one to discover—virtually, utilizing computer software—the behavior of a material by fully stretching it. The result is an object whose carbon fiber skin appears to have been sucked in, thus creating a multitude of folds. Ross Lovegrove has a keen interest in the convergence of material and form. This carbon bench, produced in total of twelve examples, should not be confused with the one made in aluminum in 2002, which was intended to be a hybrid between a solid and a liquid.

MV

Vertebrae

2004
Sculpture: Carbon fiber on a digitally milled foam core; 252 × 205 × 63 cm
Edition of 3 and 1 artist's proof

Vertebrae owes its origins to a dolphin's spinal vertebra found on Mindanao Island in the Philippines. The vertebra was scanned in 3-D and then digitized. It was given a second life molded in carbon fiber on a core of milled foam, but on an entirely different scale because it measures more than two meters high. This oversized scale, combined with the carbon fiber, allows us to look at its surface in a new way. *Vertebrae* thus encourages another level of understanding of biological evolution, beyond a "too human" approach.

MV

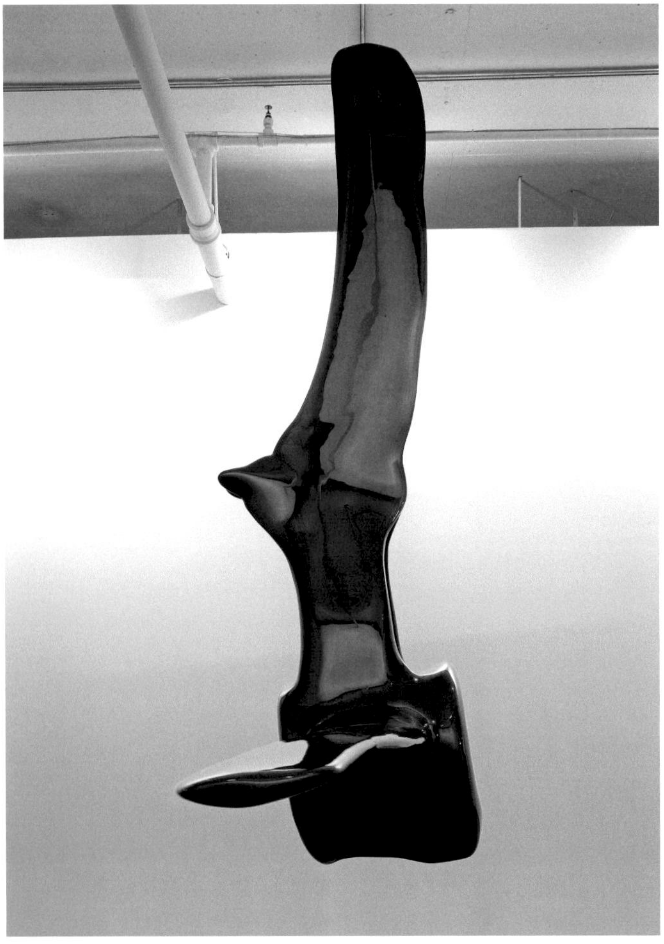

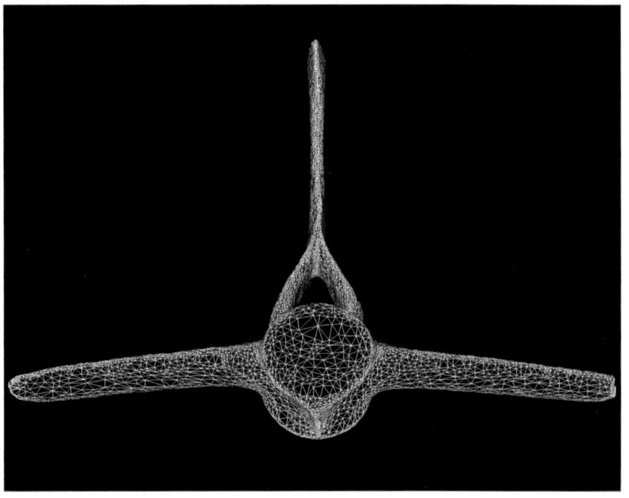

Novartis staircase

2009
Kevlar
Manufacturer: Novartis Pharma AG (Switzerland)
Novartis Campus, Basel
Architect: David Chipperfield

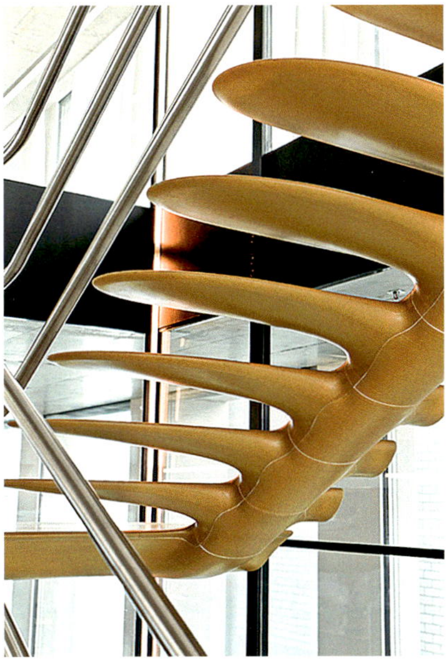

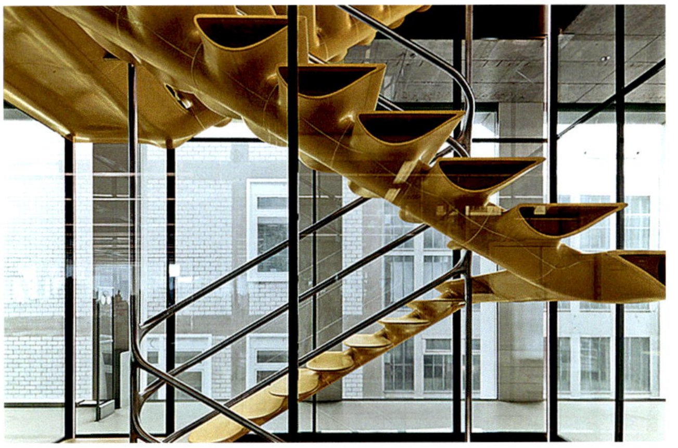

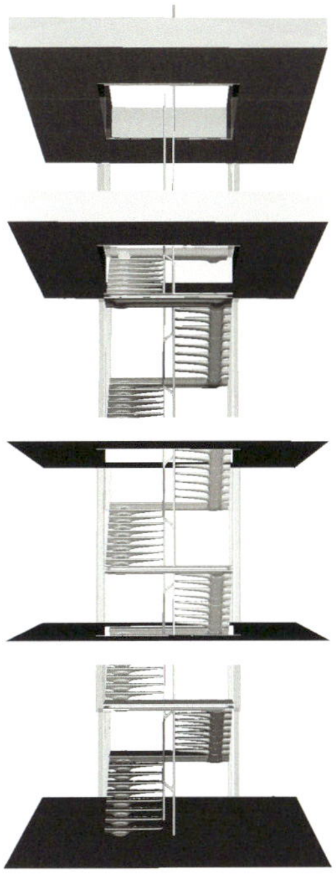

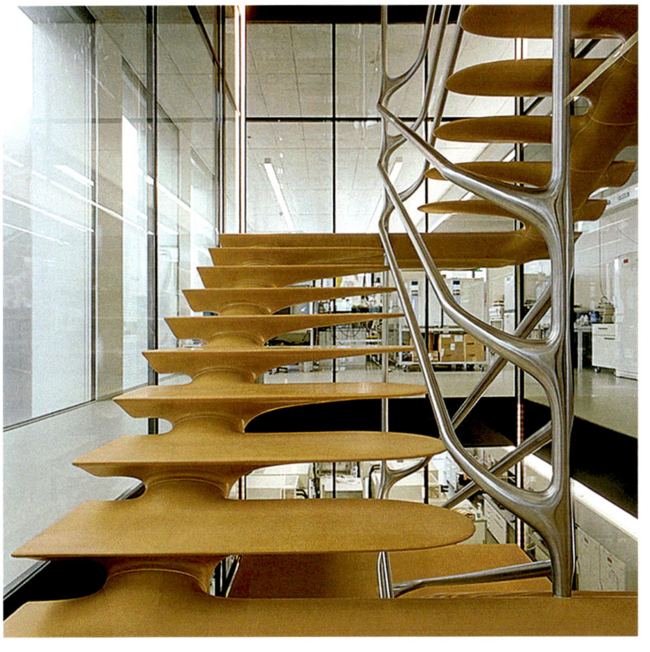

Disc Camera

1982–1983
Prototype
Acrylic; 25 × 8 × 3 cm

Disc Camera is the result of work completed at the end of Ross Lovegrove's studies at the Royal College of Art in London. It was then manufactured by Kodak. The camera works using a system of microformat rotating discs. The aim of this project was to create a new system of photographic formatting. Thanks to detailed research, Lovegrove succeeded in reducing the weight of the manufactured camera to that of a beverage can. The aluminum used for the cover and the polycarbonate chosen for the frame contributed to the lightness of the object. The photographic film itself is packed into a cassette with a ring lever attached.

MV

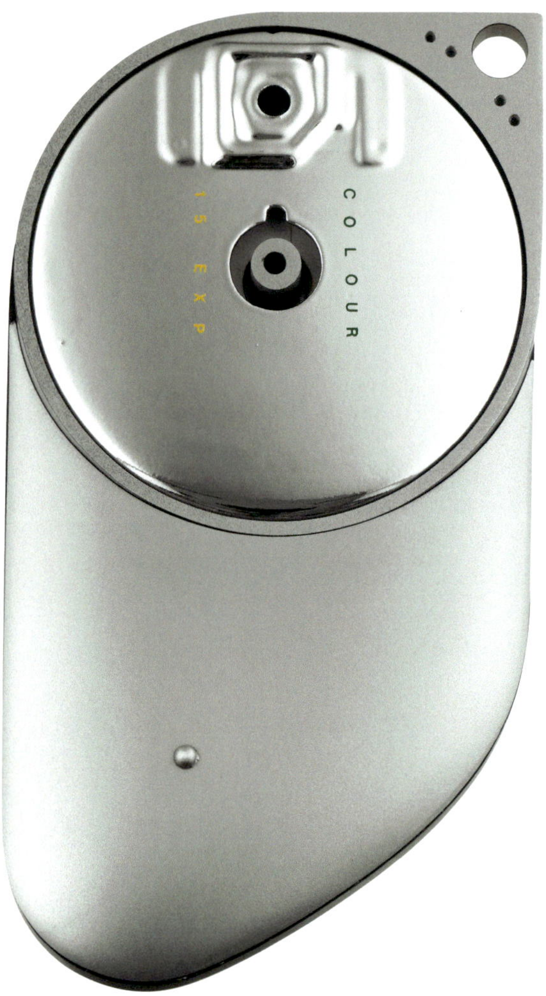

Hu watch

2005–2008
Titanium; 10 × 5 × 10 cm
Manufacturer: Issey Miyake Watch (Japan)

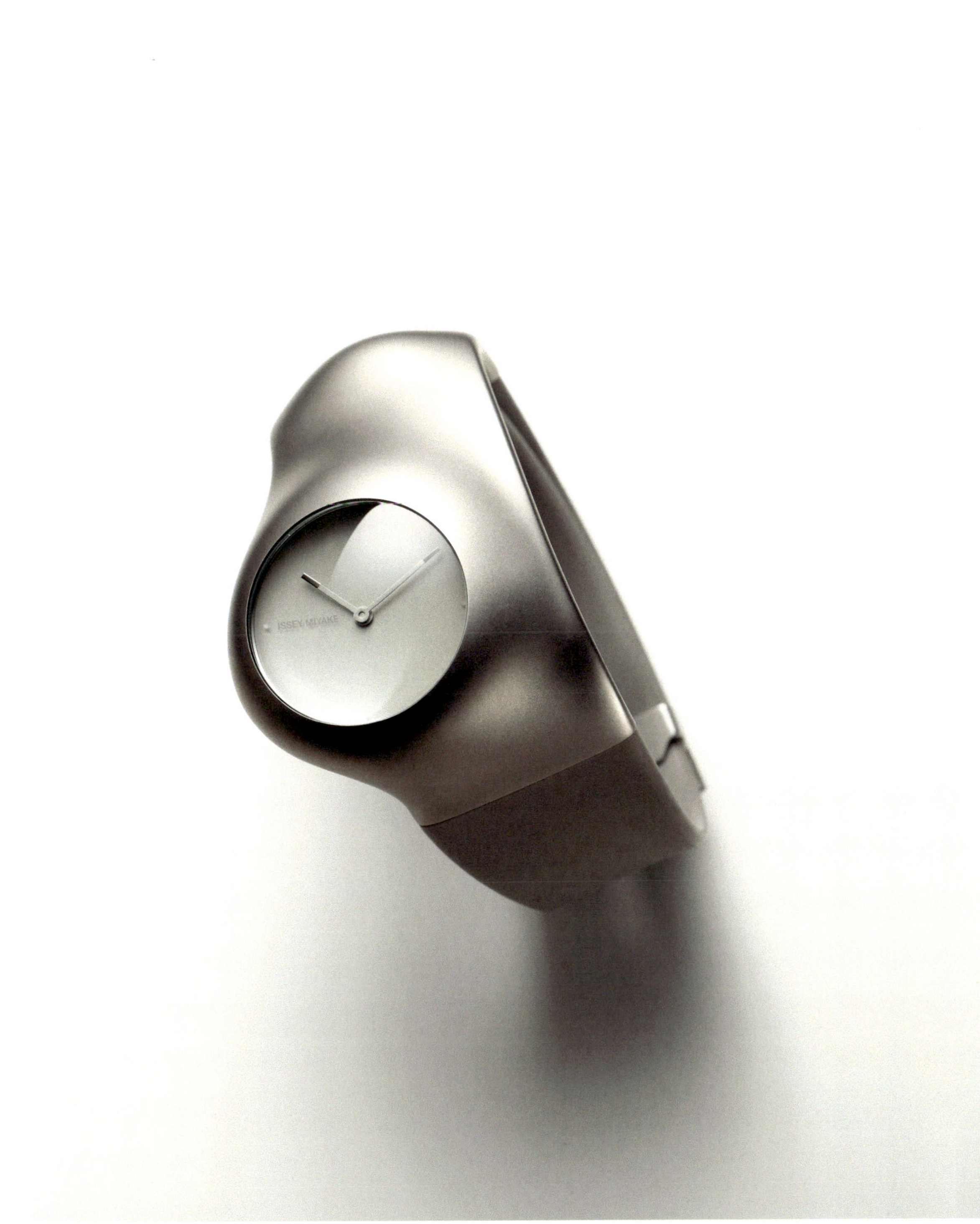

Muon speakers

2005–2007
Superformed aluminum; 200 × 38 × 60 cm
Manufacturer: KEF (United Kingdom)

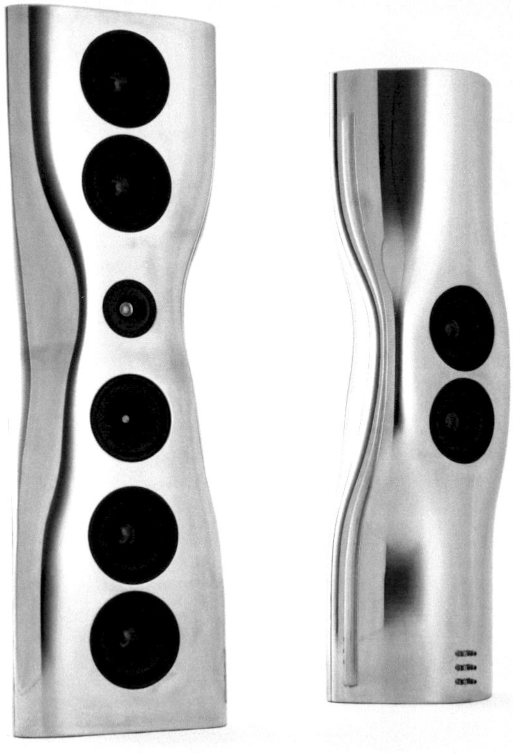

These monumental, two-meter-high loudspeakers, produced by the British firm KEF in an edition of just one hundred, were conceived as architectural elements designed to make forms in space and dynamic sound converge. With a gleaming surface produced by a 6mm-thick layer of aluminum, their sinuous appearance underscores this convergence and generates, according to Ross Lovegrove, a "fluid experience" suggesting the impalpable acoustic "material" that they produce. The word "Muon" originally derives from a Zen Buddhist term signifying a vanished continent, a question without reply. With its four-way high speaker system benefitting from the Uni-Q® system developed by KEF, which achieves a homogeneous sound without any distortion in the listening environment, Muon is a unique architectural and acoustic object.

MV

Megabioform table

2000
Aluminum; 72 × 290 × 162.5 cm
Edition of 6 and 2 artist's proofs

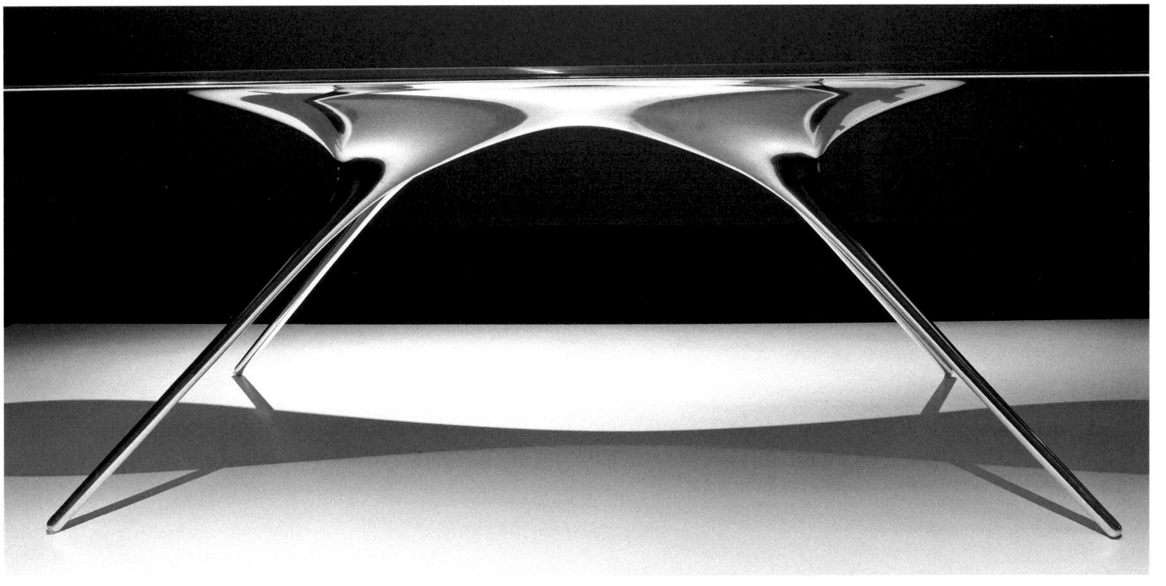

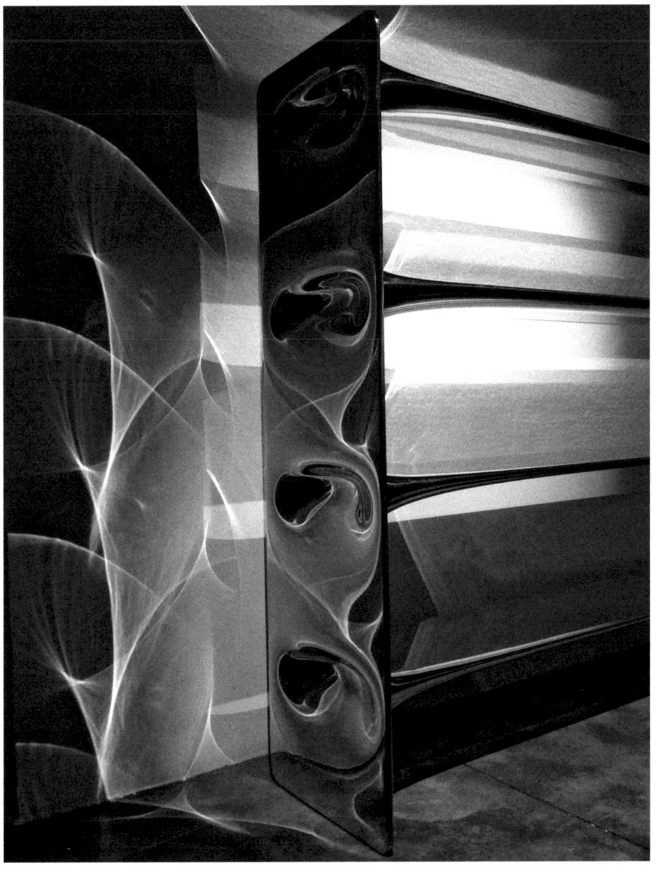

Liquid Shelves

2011
Aluminum; 185 × 300 × 50 cm
Edition of 4 and 2 artist's proofs

Eye Digital Camera

1990
Elastomer, rubber; 15 × 8 × 3 cm
Manufacturer: Olympus (Japan)

The first digital camera molded in elastomer and rubber, Eye Digital Camera embodies the change from analogue film to the memory card. The buttons usually found on electronic cameras are here replaced by various internals components. The camera can be connected to a computer or a television by a cable. The structure of the object was inspired by the prototype cameras for children made by Achille and Pierre Castiglioni for the Ferrania company in 1958, as well as the Divisumma 18 calculator, designed by Mario Bellini in 1972 and then manufactured by Olivetti. The result is an organic and biomorphic form, with the external components and the hydroelastic skin completing the outside covering of the camera. Ross Lovegrove compares this object to a prosthesis or a "biological" device. It was during his youth that the designer developed his interest in photography and the natural materials such as wood, paper, copper, and leather that made up the camera production process: "I always followed the development of cameras in order to understand the evolution of the use of materials, from wood to plastic via metal." Lovegrove insists on the rapid and considerable metamorphoses of cameras, which, as he sees it, testify to an extraordinary level of universal progress and innovation. The biomorphism of the Eye Digital Camera, its structure in elastomer, and its manual system, compared to the integrated design of smartphones today, could be interpreted as an early application of biomimetics to everyday objects.

MV

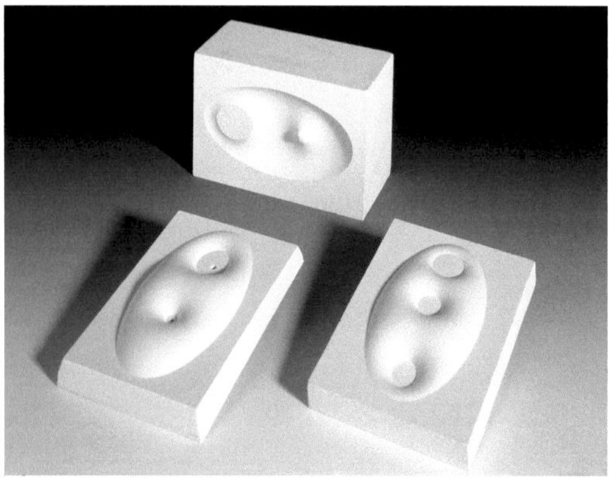

Models of resin molds;
various dimensions

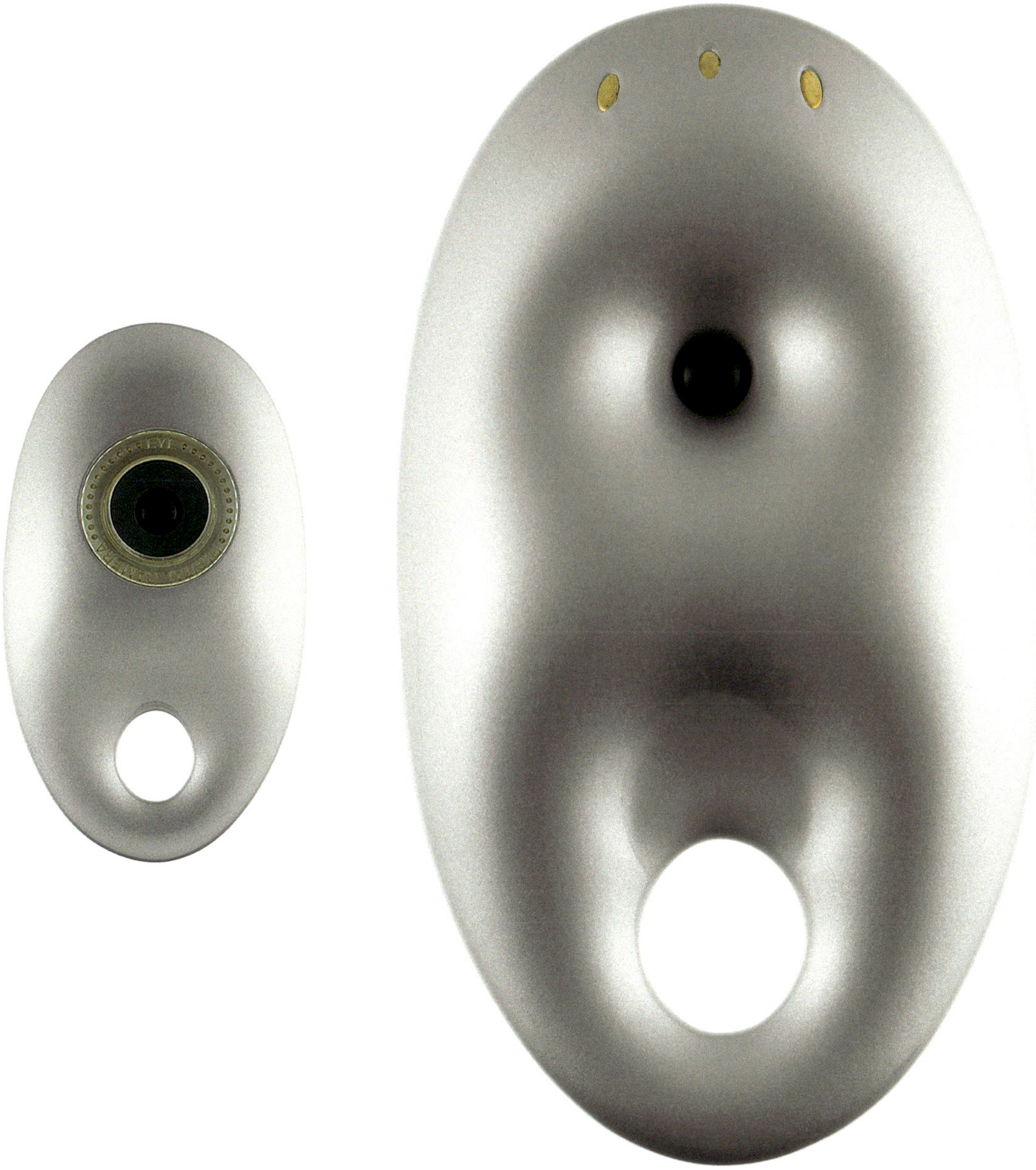

Emma Office System

1994–2000
Unique piece
Below left:
Corner table: polymer, aluminum, galvanized steel; 75 × 80 × 80 cm
Below right:
Table: polymer, aluminum, galvanized steel; ca. 70 × 190 × 45 cm
Manufacturer: Herman Miller (United States)

This office furniture system was developed by Ross Lovegrove and the furniture manufacturer Herman Miller in collaboration with the British designer Stephen Peart. It offers a new way of thinking about the relationship between the computer, the work environment, and health. A semitransparent worktop in polymer, with a honeycomb design, conceals numerous electronic cables with the help of a unique tiling system comprised of modules of pressed galvanized steel. The base and the structure of the worktable are made with compressed molded aluminum. In addition to providing a uniform work surface, the system innovates by optimizing the flow of light, which is not only beneficial for the mood and health of the worker but also lends the project a multisensorial dimension.
MV

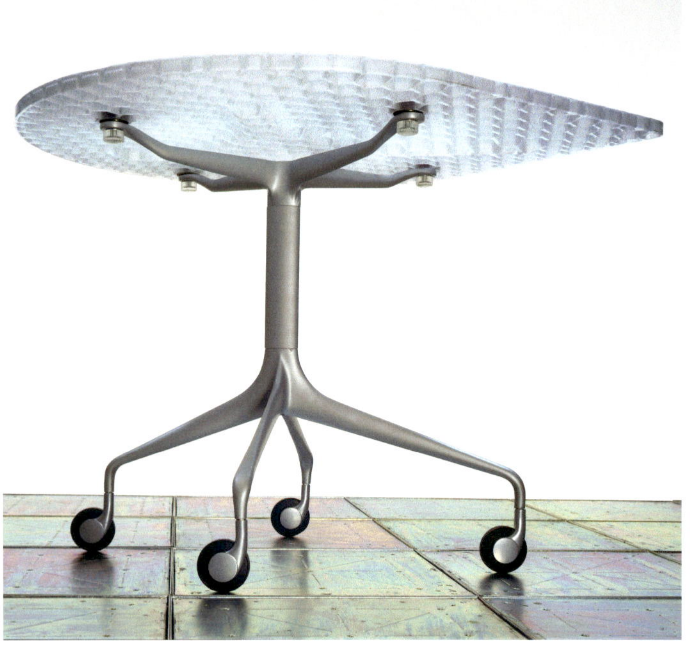

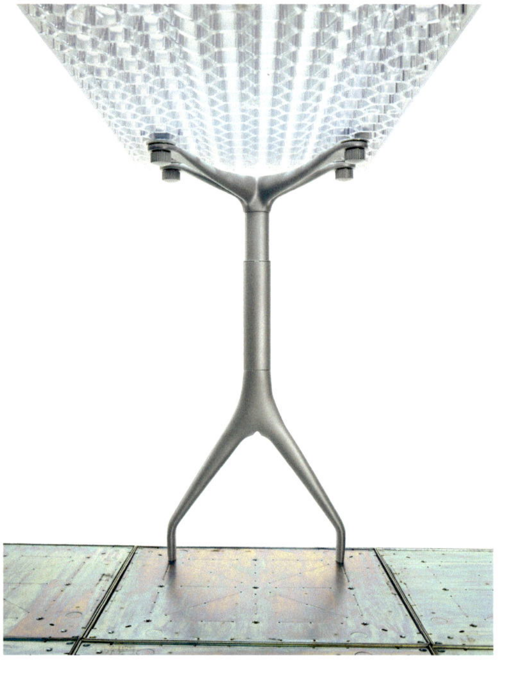

Airplane seats for JAL

1998–2002
Project for 2,000 seats for the Japan Airlines fleet
Scale model 1:20: mixed materials; 60 × 110 × 60 cm
Manufacturer: Japan Airline (Japan)

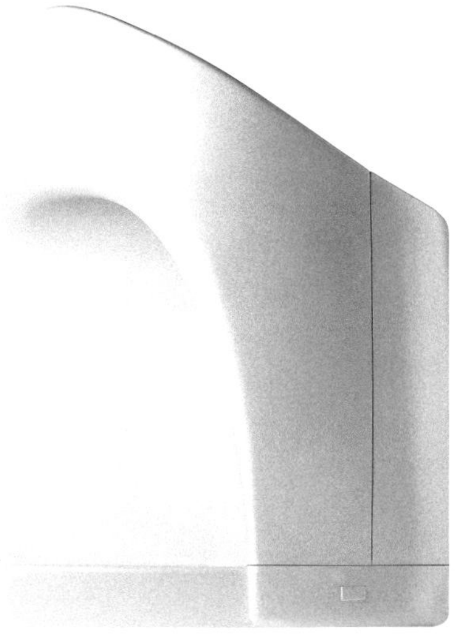

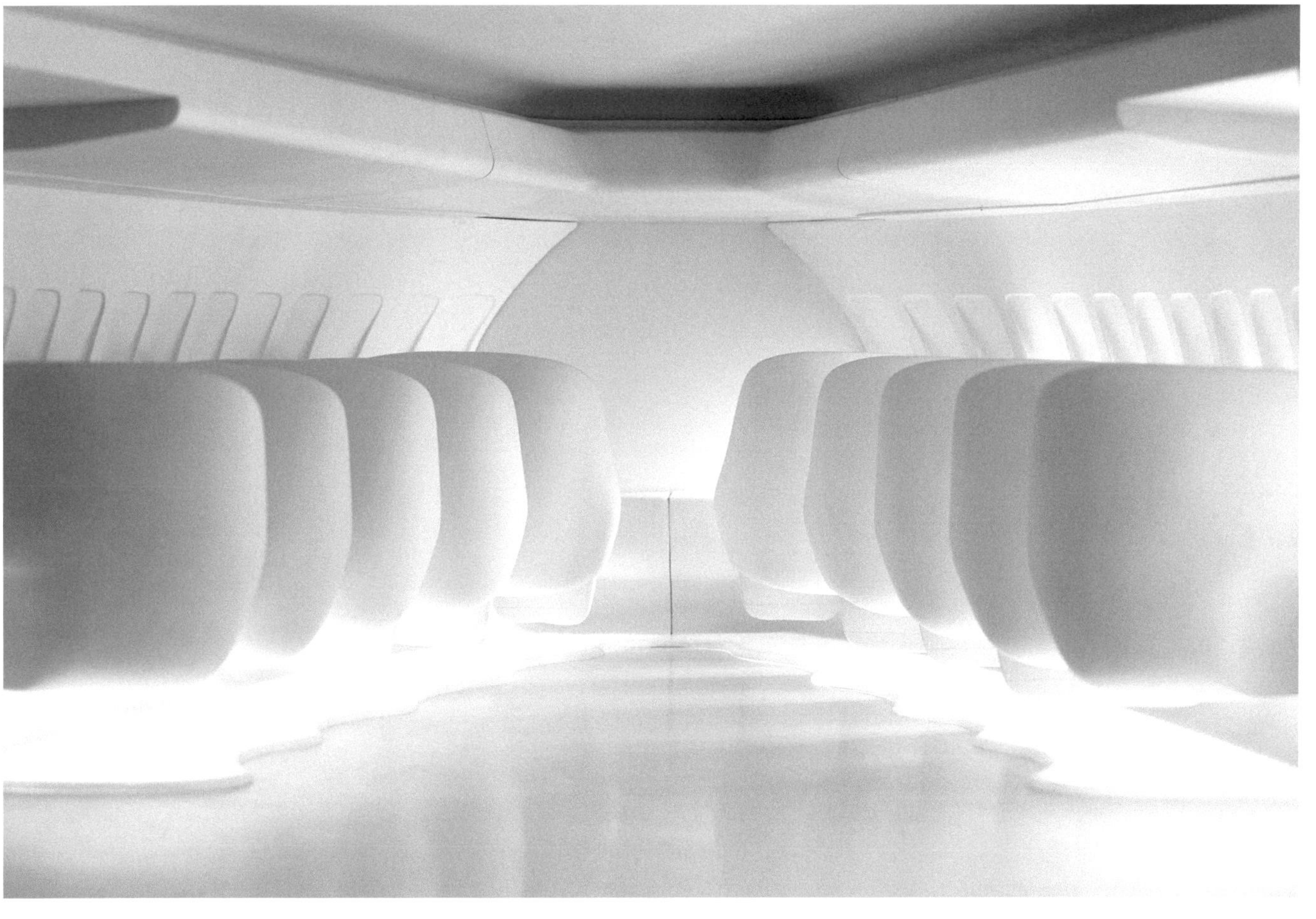

Tŷ Nant mineral water bottle

1999–2001
Polyethylene terephthalate (PET); H.: 23 cm, Dia.: 7 cm
Manufacturer: Tŷ Nant (United Kingdom)

In 1999 the Welsh mineral water company Tŷ Nant commissioned Ross Lovegrove to design a completely new form of plastic bottle. To achieve this, Lovegrove thought about designing a bottle that went beyond a simple industrial product. Indeed, he wanted the object's form to reflect its contents. The bottle was thus the outcome of an industrialization process that lasted more than a year. The challenge that the designer set himself for this project was to create a surface that captured the essence of water, its purity, fluidity, and elusive character, while devising a form that gave the impression of having captured its movement. The result is an asymmetrically shaped bottle made of polyethylene terephthalate (PET) and produced by computer. The Tŷ Nant bottle is the first industrial product in the world to have been made employing the algorithms of the Nonuniform rational B-spline (NURBS) modeling system used in parametric architecture, in order to obtain a perfect union between its material, its technology, and its anatomy.
Six Tŷ Nant bottle shapes exist, produced in three formats (50, 100, and 150 cl), made in separated aluminum units to allow the necessary production volumes. Thus, the Tŷ Nant bottle, despite its multiple hollows and cavities, bears unique ergonomics that makes it very easy to handle. Its sculptural elegance—rare for a basic consumer product—opens up new creative horizons for the world of industrial packaging.

MV

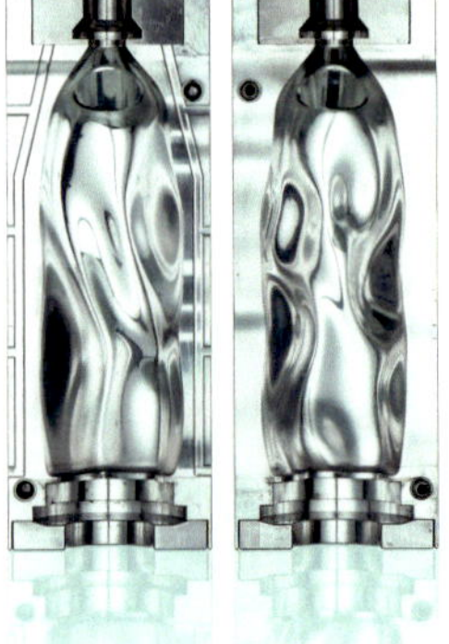

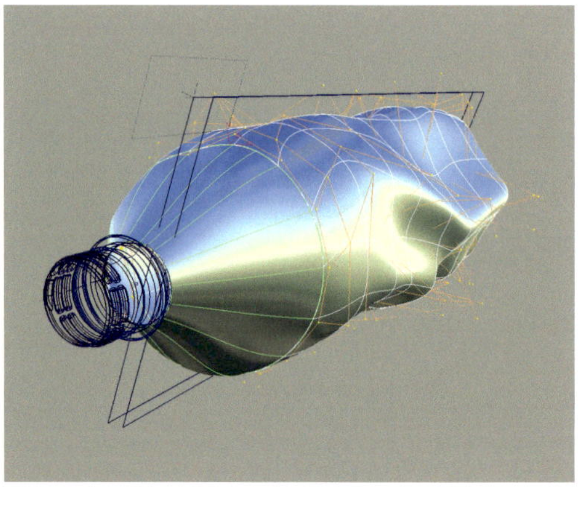

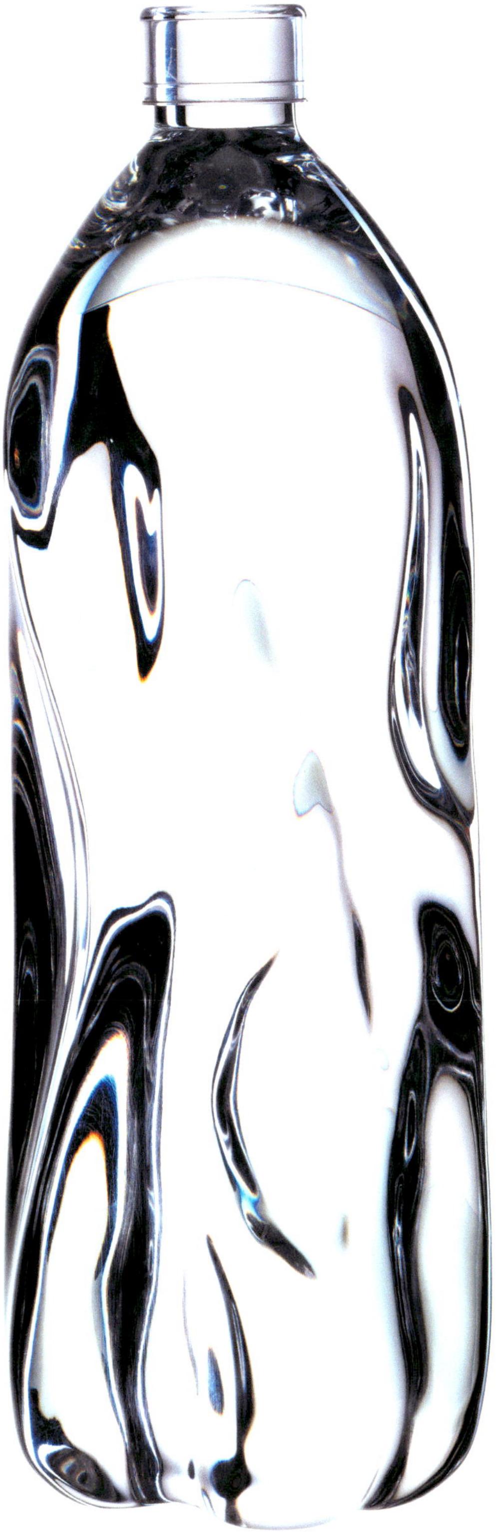

Opposite, from left to right:
Aluminum molds;
50 × 25 × 25 cm
Computer-generated image
Sketch made on a digital tablet
Right:
Prototype in polythene;
H.: 23, Dia.: 7 cm
Ross Lovegrove Collection

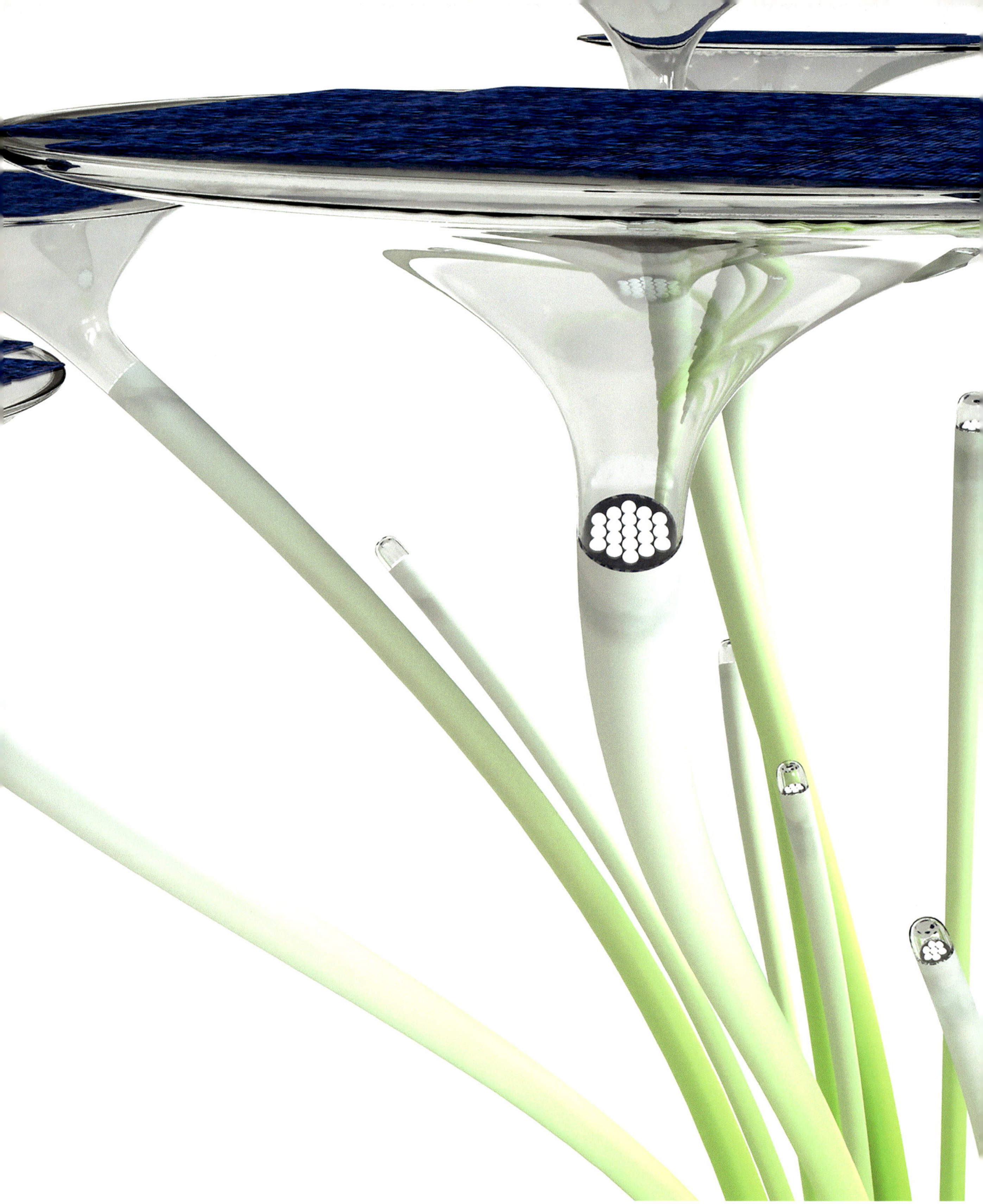

Solar Tree lamp, 2006 – 2010 (detail),

See pp. 86 – 87

Eco

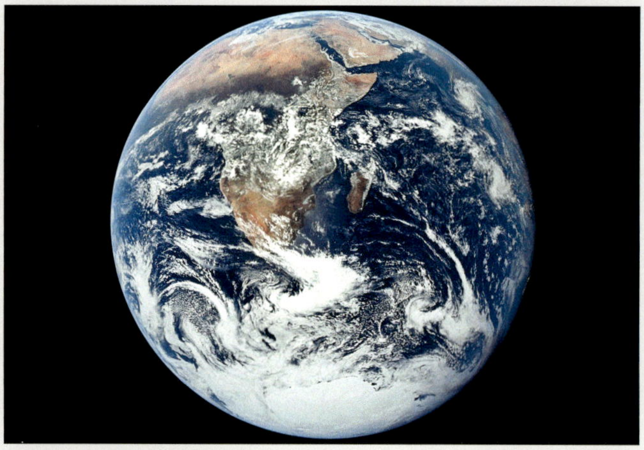
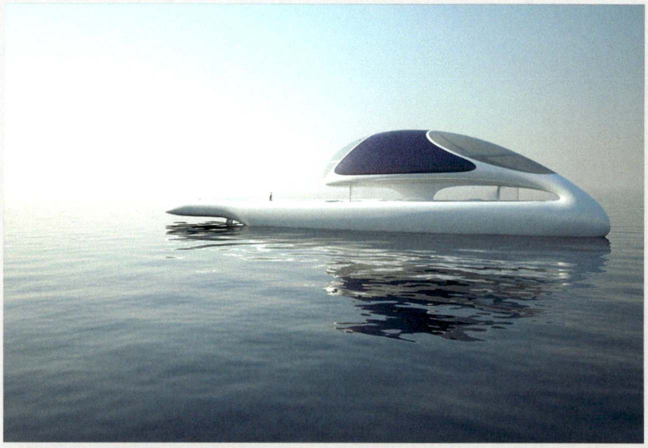
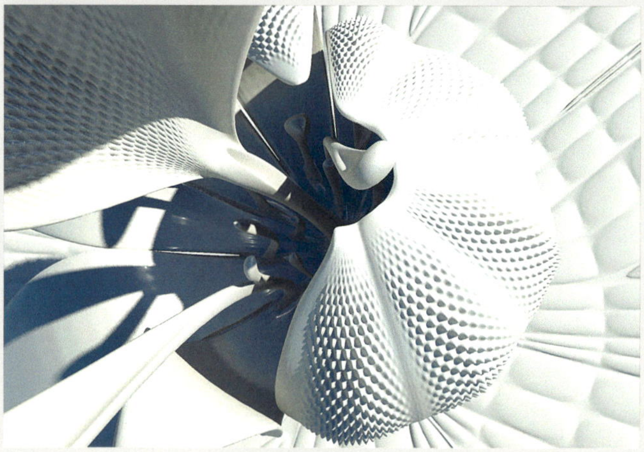
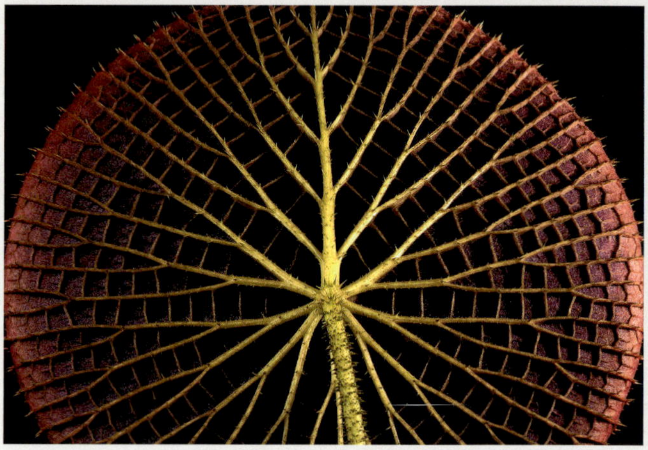
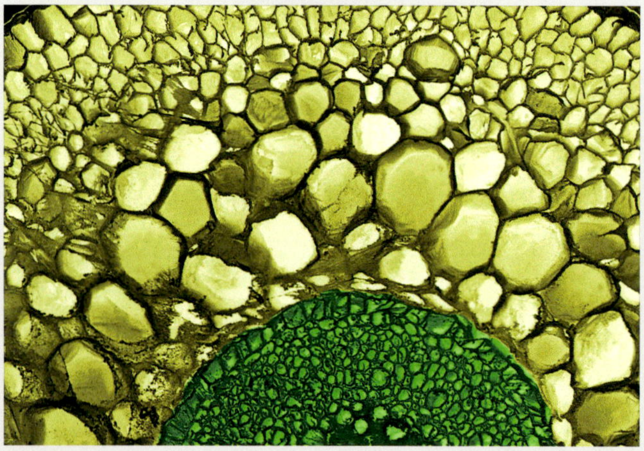
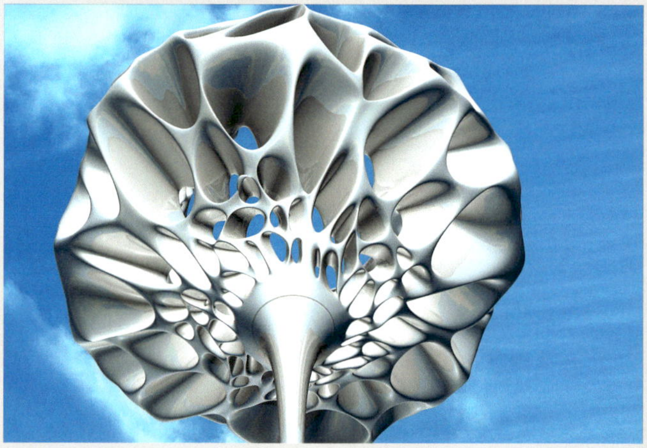

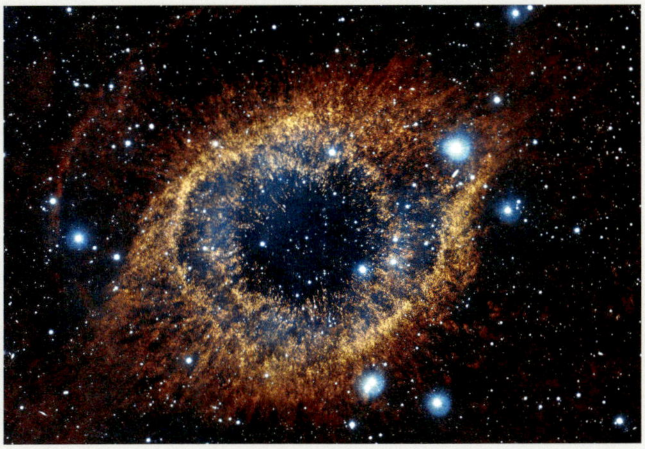

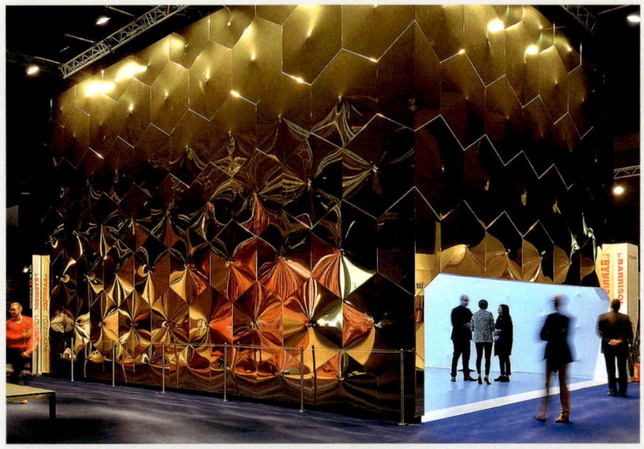

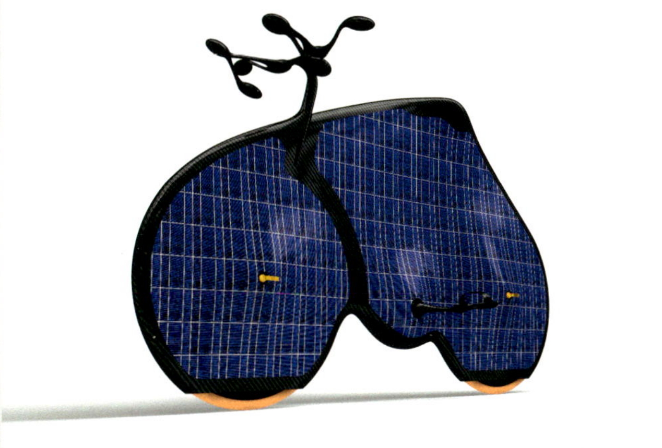

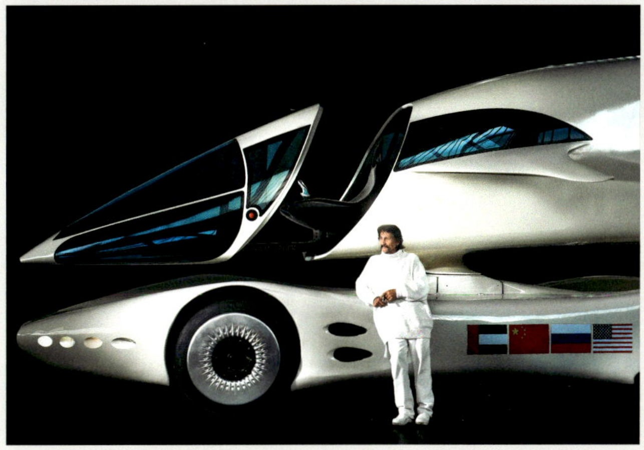

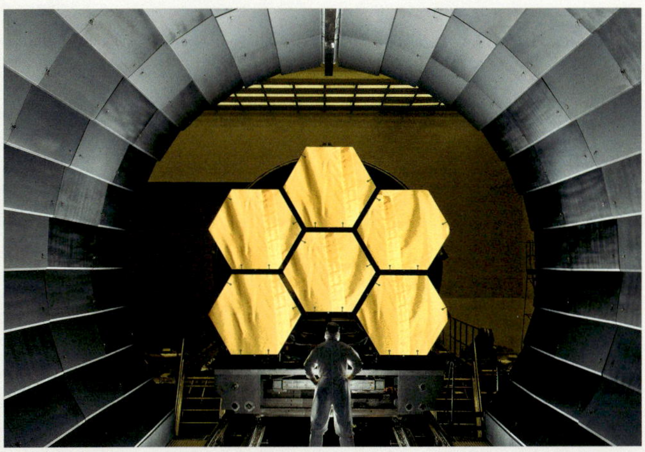

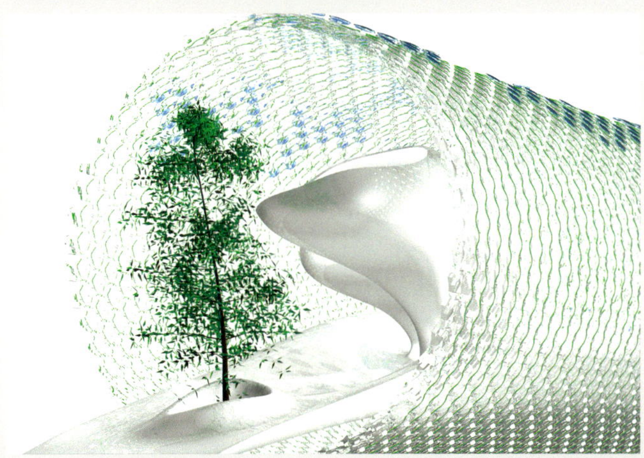

Page 82 (*from left to right and top to bottom*)

Earth photographed by the Apollo 17 astronauts, traveling toward the moon, December 7, 1972

Ross Lovegrove, Bioyot, 2008, project for a yacht partially powered by solar energy, computer-generated image, L.: 60 m, Ross Lovegrove Collection

Ross Lovegrove, *Calgat* [Can a Lily Grow a Telephone] (detail), 2008, 3-D printed sculpture, 80 × 80 × 80 cm, Ross Lovegrove Collection

Jonathan Singer, *Underside of Leaf, Giant Amazon Water Lily (Victoria Amazonica Sowerby)*, chromogenic proof, 2008

Cross section of the root of an epiphytic orchid seen through a microscope, showing the water-storage tubes

Ross Lovegrove, Floral Gyrator (detail), 2006–2008, wind turbine project, computer-generated image, Ross Lovegrove Collection

Page 83 (*from left to right and top to bottom*)

View of the Helix planetary nebula (NGC 7293), situated 700 light years from Earth, captured by the Visible and Infrared Survey Telescope for Astronomy (VISTA), at the European Southern Observatory (ESO), Chile

Ross Lovegrove, Barrisol Pavilion, 2014, stretched Barrisol fabric and three-dimensional tubular frame; 700 × 1200 × 1200 cm, manufacturer: Barrisol Normalu, France (view of the Interiors Biennale, Courtrai, Belgium, 2014)

Ross Lovegrove, Solar Skin, 2011, bicycle project in carbon fiber covered with photovoltaic cells, computer-generated image, Ross Lovegrove Collection

Luigi Colani, Streamline mobile home, 2006–2008

Ernie Wright in front of the hexagonal mirrors of the James Webb Space Telescope at the NASA Marshall Space Flight Center, Huntsville, Alabama, United States

Ross Lovegrove, Tube Architecture, 2002–2003, architectural project powered by solar panels, Ross Lovegrove Collection

For Ross Lovegrove, biology, anthropology, physics, and ecology are fields that fuel his humanist vision of design turned to face the future. His use of the latest technologies is tied to an environmental approach to create objects in which dynamic shapes and hybrid materials intersect. This ecological approach is a response to actual issues linked to the environment, and the designer's productions become part of his overall global thinking. Lovegrove emphasized this commitment as early as 1994, in a research project about nomadic architecture, Solar Seed, then with the Solar Bud light in 1998. In 2006, he created Solar Tree, a group of lights that engage solar energy for the creation of low-cost ecological lighting in public spaces. This pioneering research into the use of solar energy culminated in 2005–2006 at the Swarovski Crystal Aerospace with a concept car using the optical properties of Swarovski crystals to focus sunlight on the photovoltaic cells covering the body of the vehicle. At that same moment, Lovegrove created the Generator House, in collaboration with the engineering firms Carbo Tech and Arup. This concept of a futurist house marked a crucial stage in the designer's ecological and environmental work. Offering solutions adapted to climate and atmospheric changes, he designed a habitat capable of being perfectly integrated into a natural environment and hence, the completely transparent façade was equipped with photovoltaic cells. In 2012, for the Renault Twin'Z, a car that is one hundred percent electric, he used composite and recycled materials to reduce pollution emissions linked as much to the manufacturing of the product as to its actual use. Beyond the field of experimentation and the empirical approach to his collaboration with cutting-edge industry in research and development, Ross Lovegrove asserts the position of the designer in his era. The design and production of functional objects using technology has to respond to the development of new lifestyles. Thinking about the necessary transition "from the 'hard-line' mechanical aesthetic . . . to the ecological principles of a 'softer' aesthetic," Ross Lovegrove has been a pioneering figure in conceptualizing environmental and ecological ideas using technoscience, in order to apply them directly to the field of design.

Olivier Zeitoun

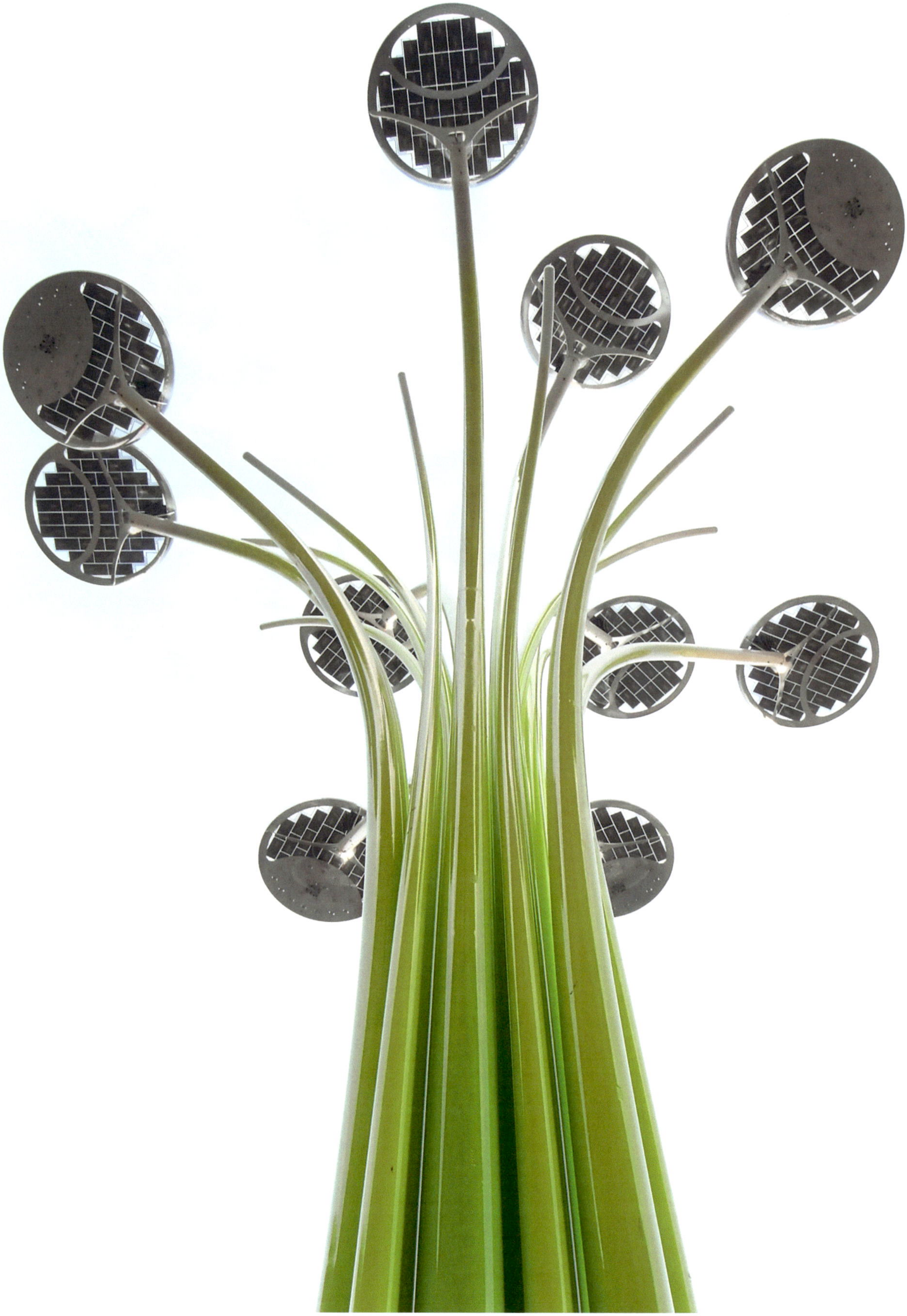

Solar Tree lamp

2006–2010
Steel, concrete, LED, photovoltaic cells;
H.: 550 cm, Dia.: 441 cm
Manufacturer: Artemide (Italy) and Ross Lovegrove

This steel light was developed in collaboration with Artemide and Sharp Solar, the market leader in solar lighting. The light that it projects comes simultaneously from a system of electroluminescent diodes (LED) and from a network of thirty-eight solar captors connected to a battery. By adopting the shape of a stylized tree, 5.5 meters high and with a circular bench at its base, the lamp offers pedestrians a different perspective on the urban environment.
It also demonstrates a particular attention to the relationship between environment, material, and technology.
The lamp's plant-like appearance is emphasized by its alternating branches, ten of which end with a one-watt LED and ten others with steel leaves projecting a light strength of 1,250 lumens. These leaf-form lamp heads have photovoltaic cells embedded in their upper surface that receive light during the day and power the LED system by charging their battery positioned in the base of the lamp. The Solar Tree can therefore function completely autonomously, but it can also be connected to the town's electrical network and send its excess energy to the central generator. Indeed, the light was designed for a hybrid usage. The batteries ensure independent functionality for seven to ten days. If it discharges completely, due to a lack of natural light, the lamp still functions thanks to its connection to the public energy system.
Beyond its ecological dimension, the Solar Tree suggests, through its symbolic form, an optimistic evocation of nature, in comparison to the typical grayness of many urban environments. At the same time, it raises inhabitants' awareness of the energy-source changes needed in the realm of everyday objects.
After its temporary installation in front of the MAK – Austrian Museum of Applied Arts in Vienna, during Design Week in October 2007, the Solar Tree was installed on other large European city squares: Potsdamer Platz in Berlin and the Piazza della Scala in Milan, as well as in Thessalonica.

MV

Solar Seed

1994–1995
Nomadic architectural project
Model: polymers; H.: 75 cm, Dia.: 65 cm

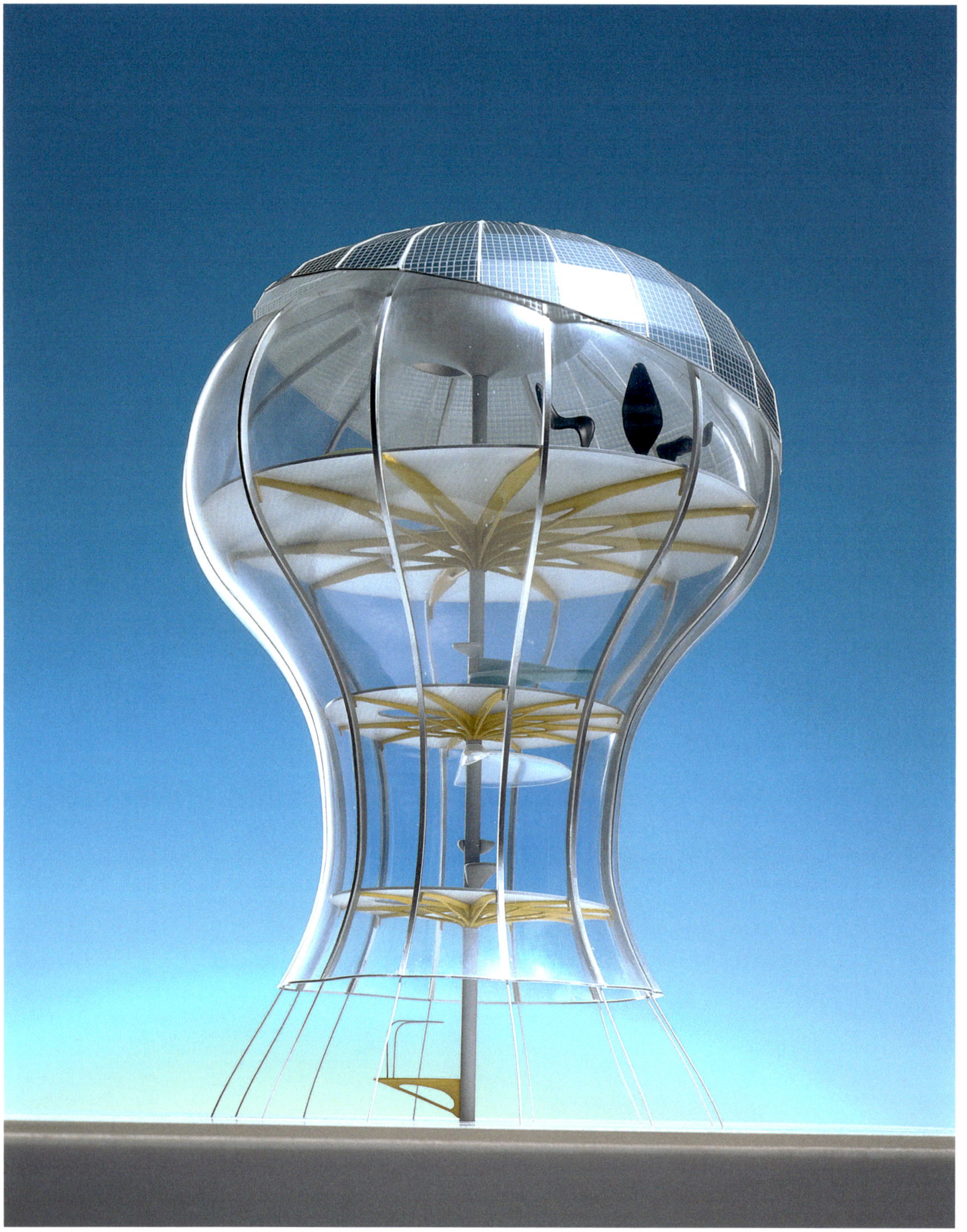

Solar Bud Light

1998
Polycarbonate, aluminum, solar-energy LED;
H.: 30 cm, Dia.: 15 cm
Manufacturer: Luceplan (Italy)

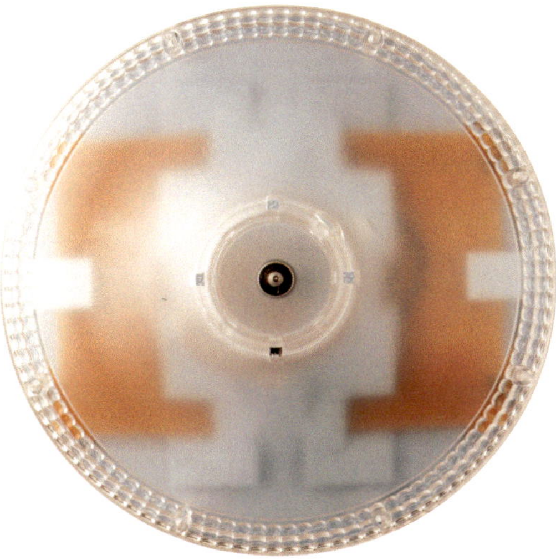

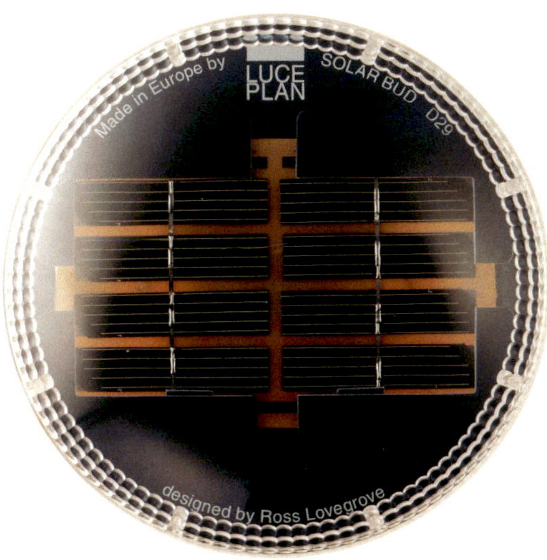

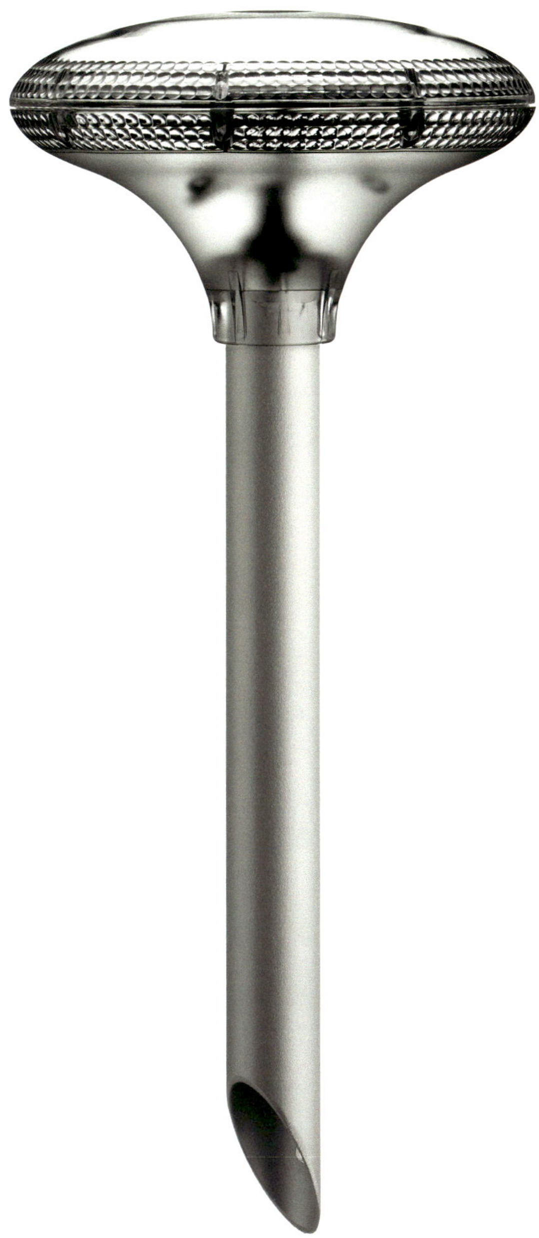

Generator House

2006–2008
Project for a self-sufficient house using solar energy
Computer-generated images

In an era of widespread social media, access to information via the Internet, and the development of off-grid technologies, more and more people aspire to live outside of towns and cities while remaining completely connected to human activity and culture.
The prospective housing concept designed by Lovegrove and realized in association with the engineering firms Carbo Tech and Arup attempts to satisfy these concrete needs and actual concerns while also getting closer to the human being, who has to be capable of adapting to climate and atmospheric change.
The house is situated in a natural environment, allowing the families occupying it to live in a healthy space, enhancing their sense of well-being and development of a certain synergy with nature. The overall structure of the house is designed using polymers and other light composite materials. The completely transparent façade is equipped with photovoltaic cells. The energy harnessed by the house is thus redistributed thanks to a turbine integrated into its architecture.
In line with the Solar Car or the Solar Tree, this project for a responsible house marks a new stage in the evolution of Lovegrove's ecological thinking.

MV

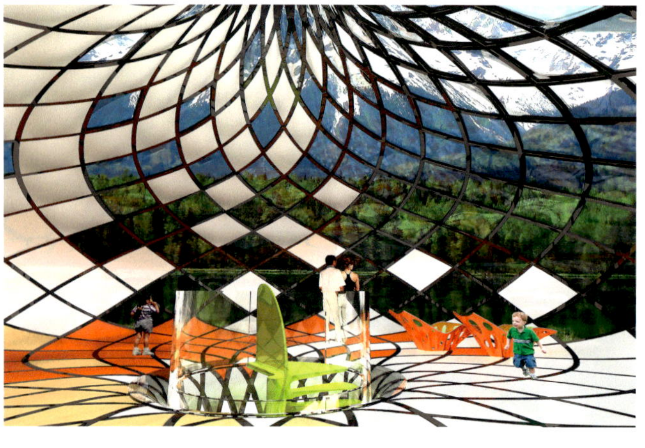

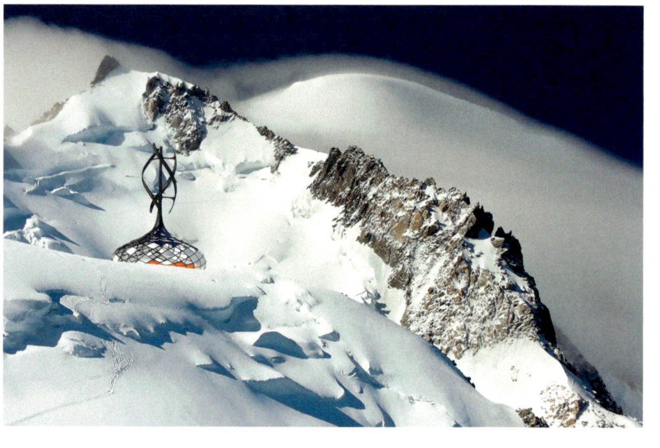

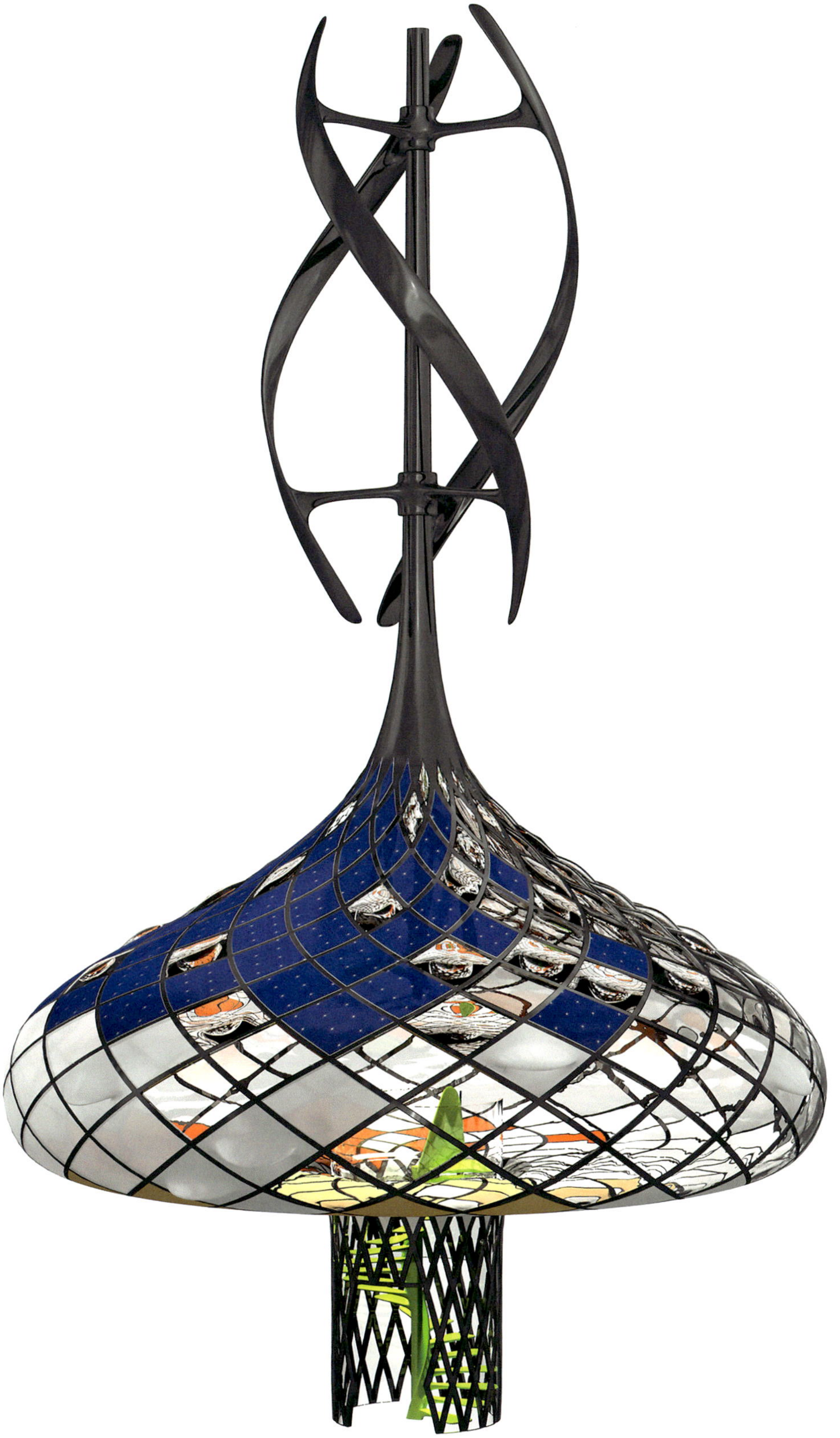

Bamboo Bicycle

2001–2008
Bamboo, aluminum; 102 × 156 × 57 cm
Manufacturer: Biomega (Denmark)

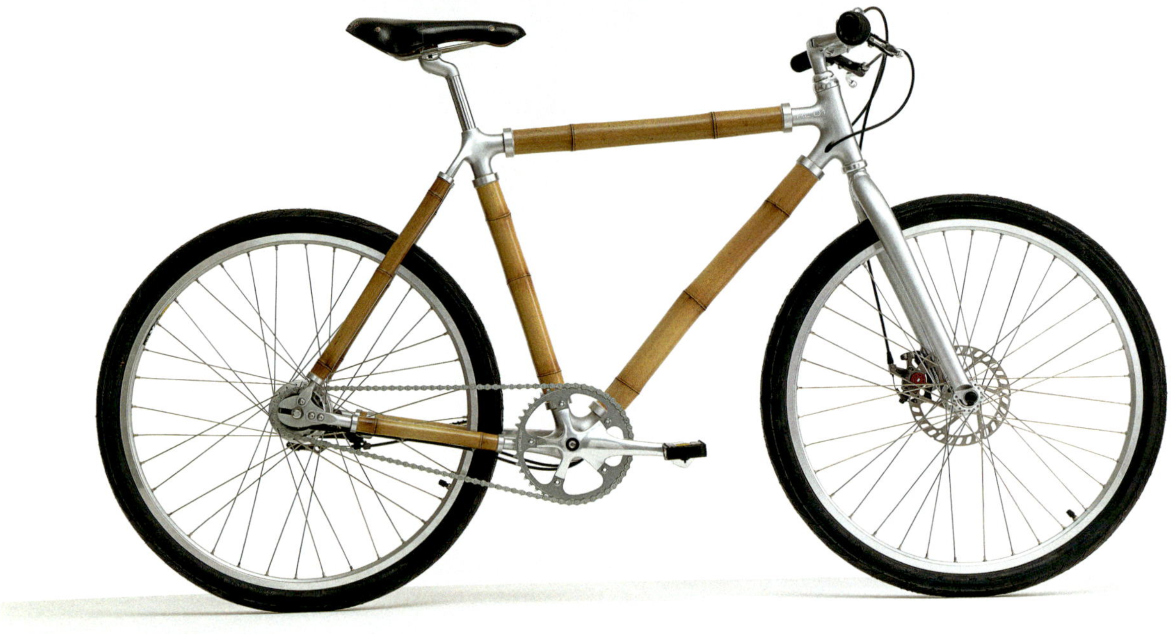

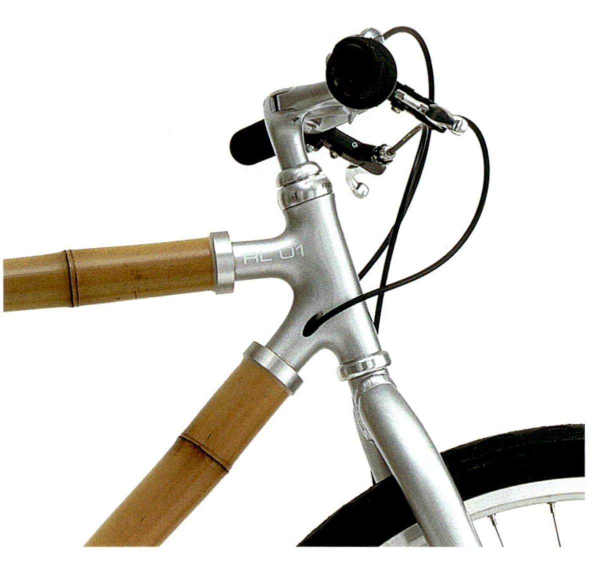

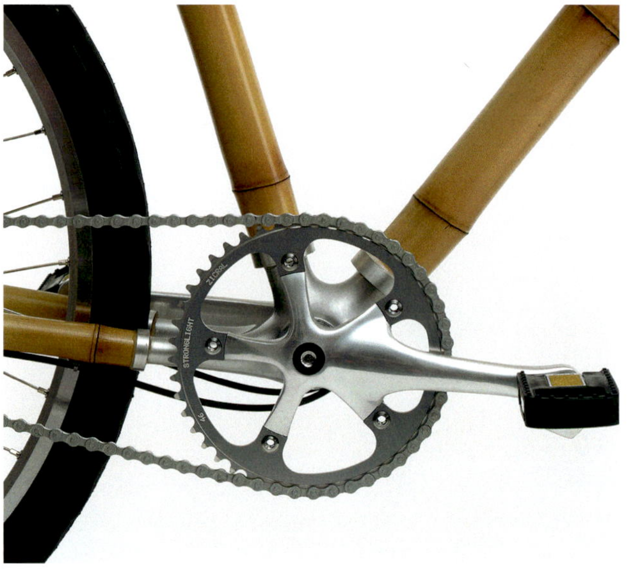

Calgat

[Can a Lily Grow a Telephone]
2008
Computer-generated image

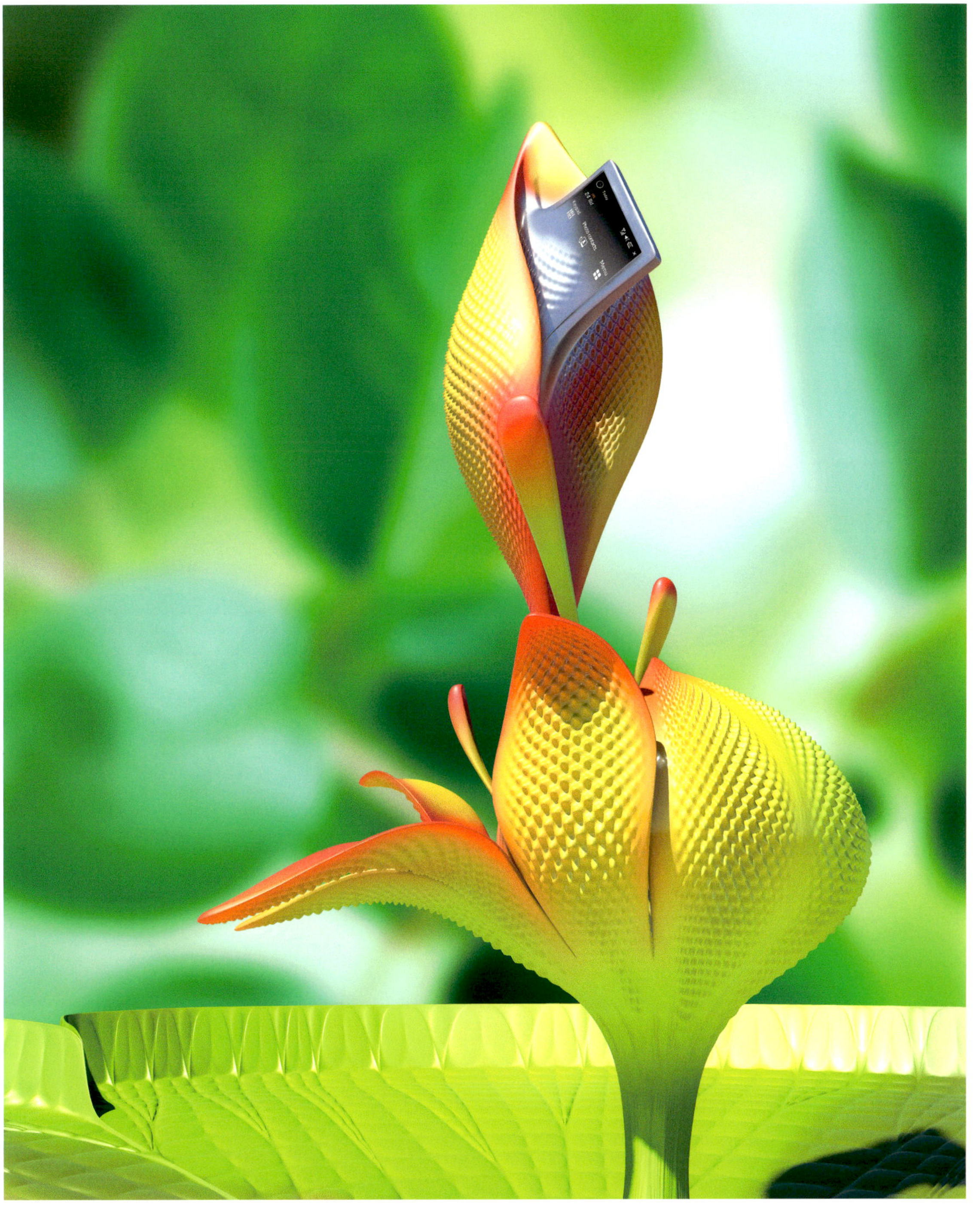

Car on a Stick

2002–2007
Urban ecological transport project
Model; mixed materials; H.: 60 cm, Dia.: 80 cm

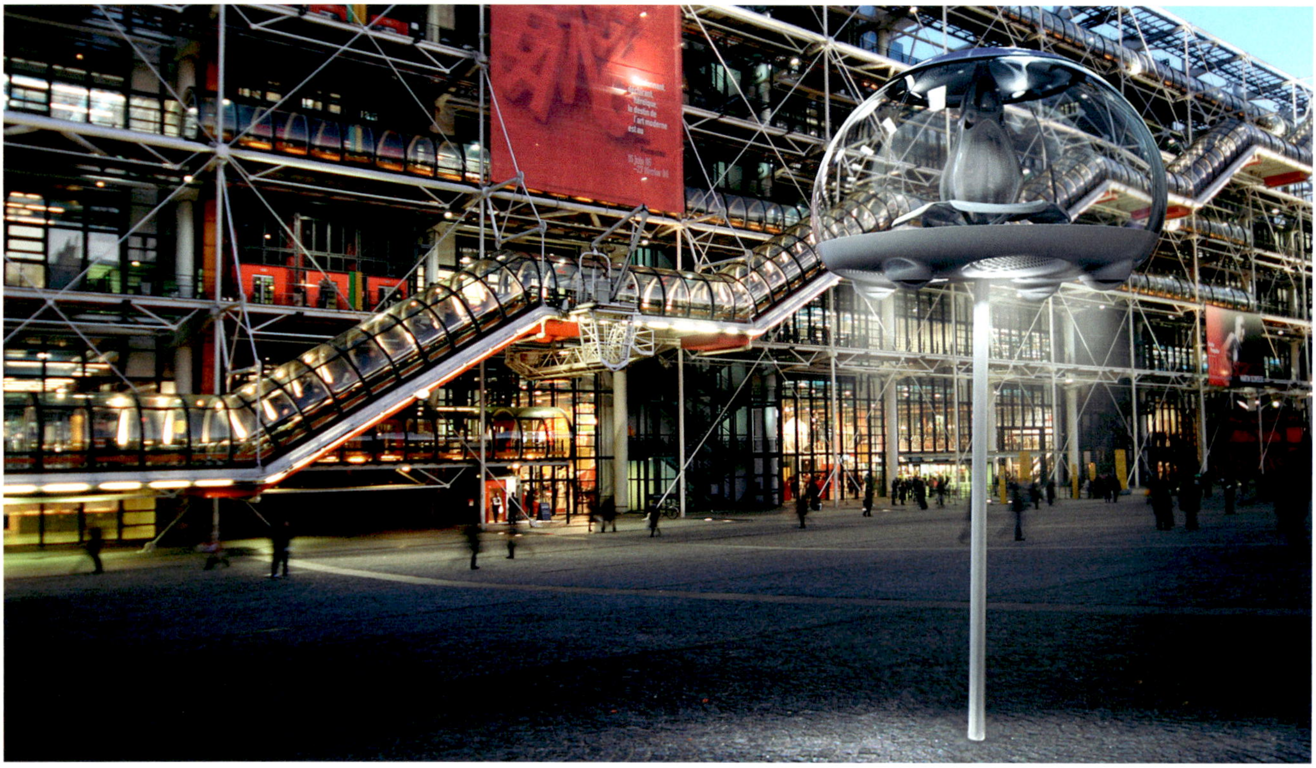

Car on a Stick is the name given to a new urban transport concept that aims at liberating towns and cities from automobiles. In this project, the cars are electric and autonomous and their energy is derived from solar panels installed on their roofs. When not in use, the bubble-shaped vehicles are "parked" at the top of telescopic poles; as night falls they transform themselves into streetlights. The streets are thus freed up and the traffic flow becomes more fluid. The vehicles are designed to carry four passengers with luggage, which can be stored in compartments integrated in the floor. The cars move automatically using voice control and advanced satellite navigation, without conventional steering. Proximity sensors facilitate navigation, in a manner that optimizes traffic flow.

MV

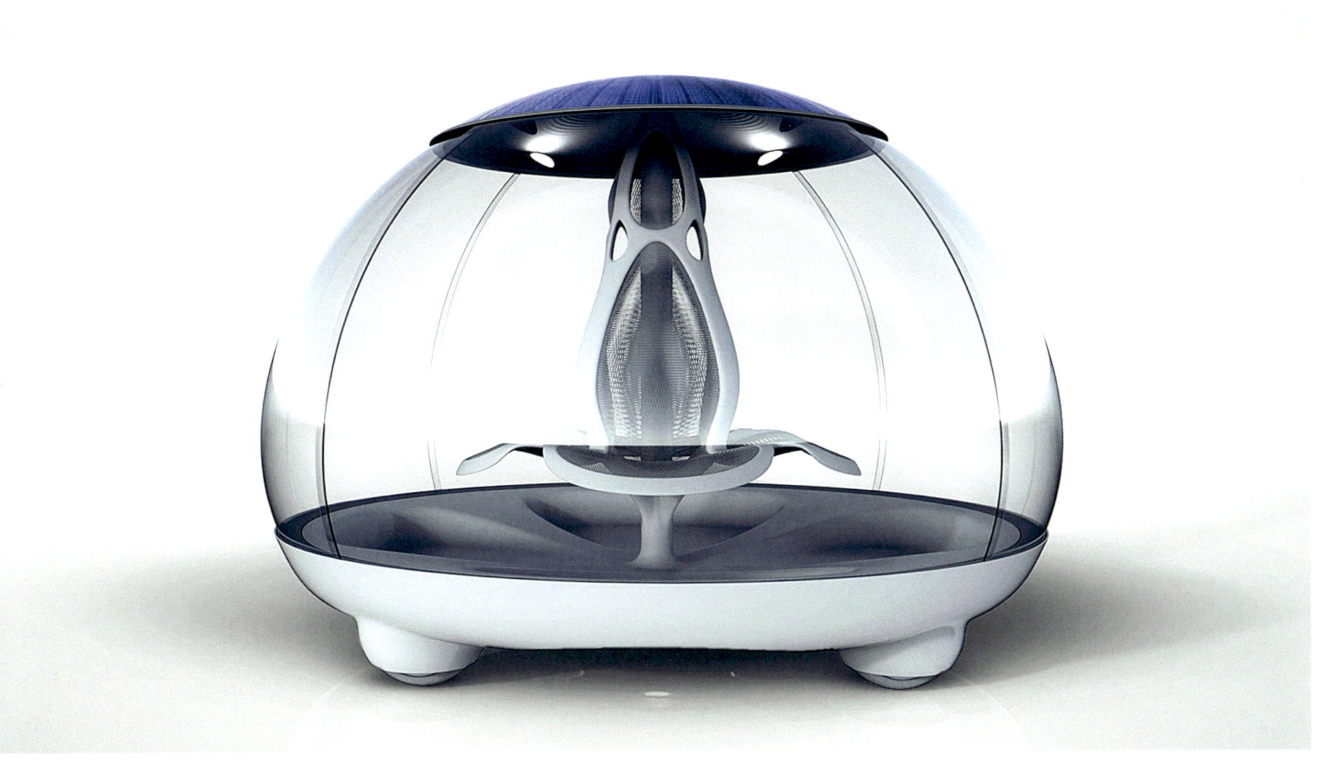

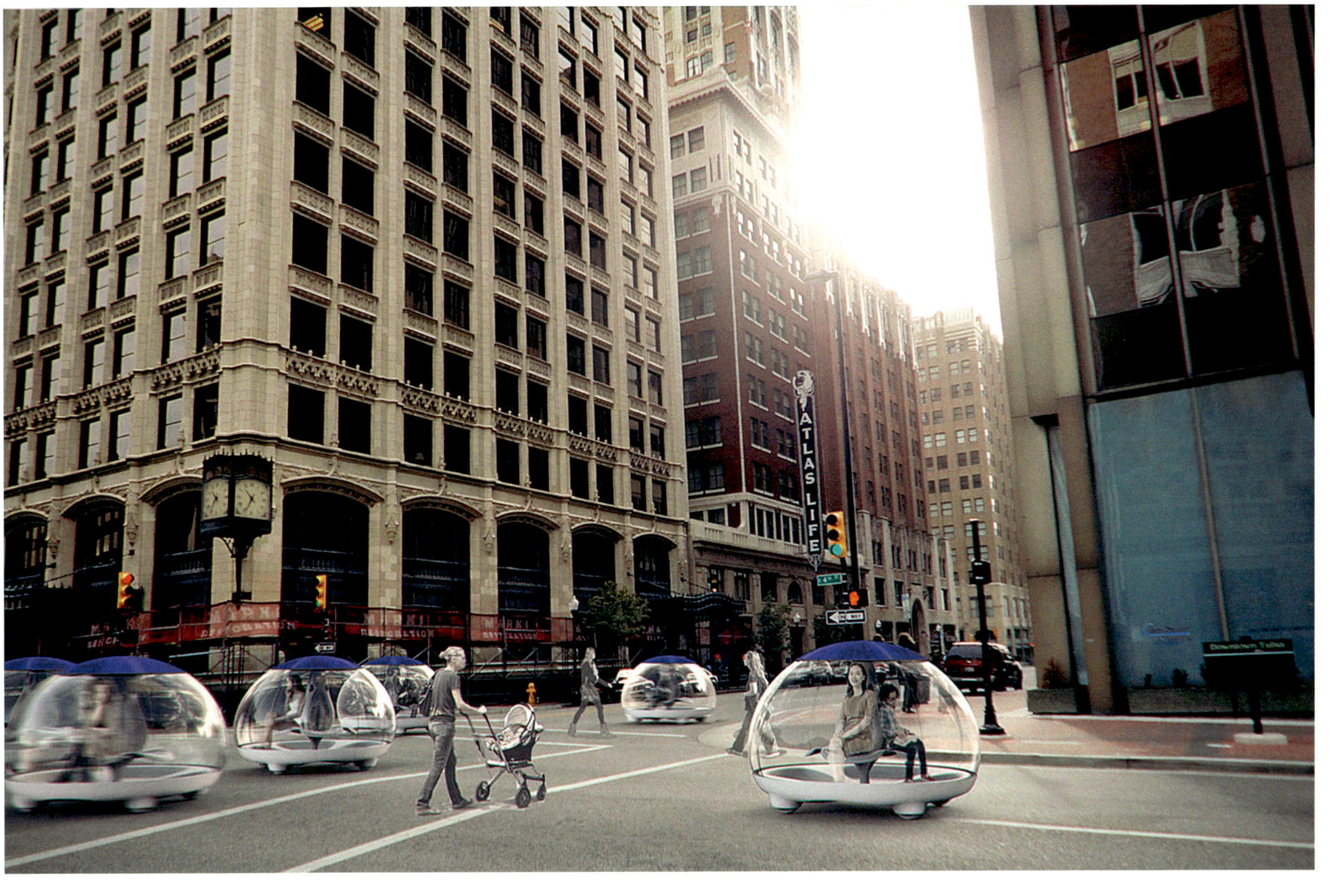
ATLAS LIFE

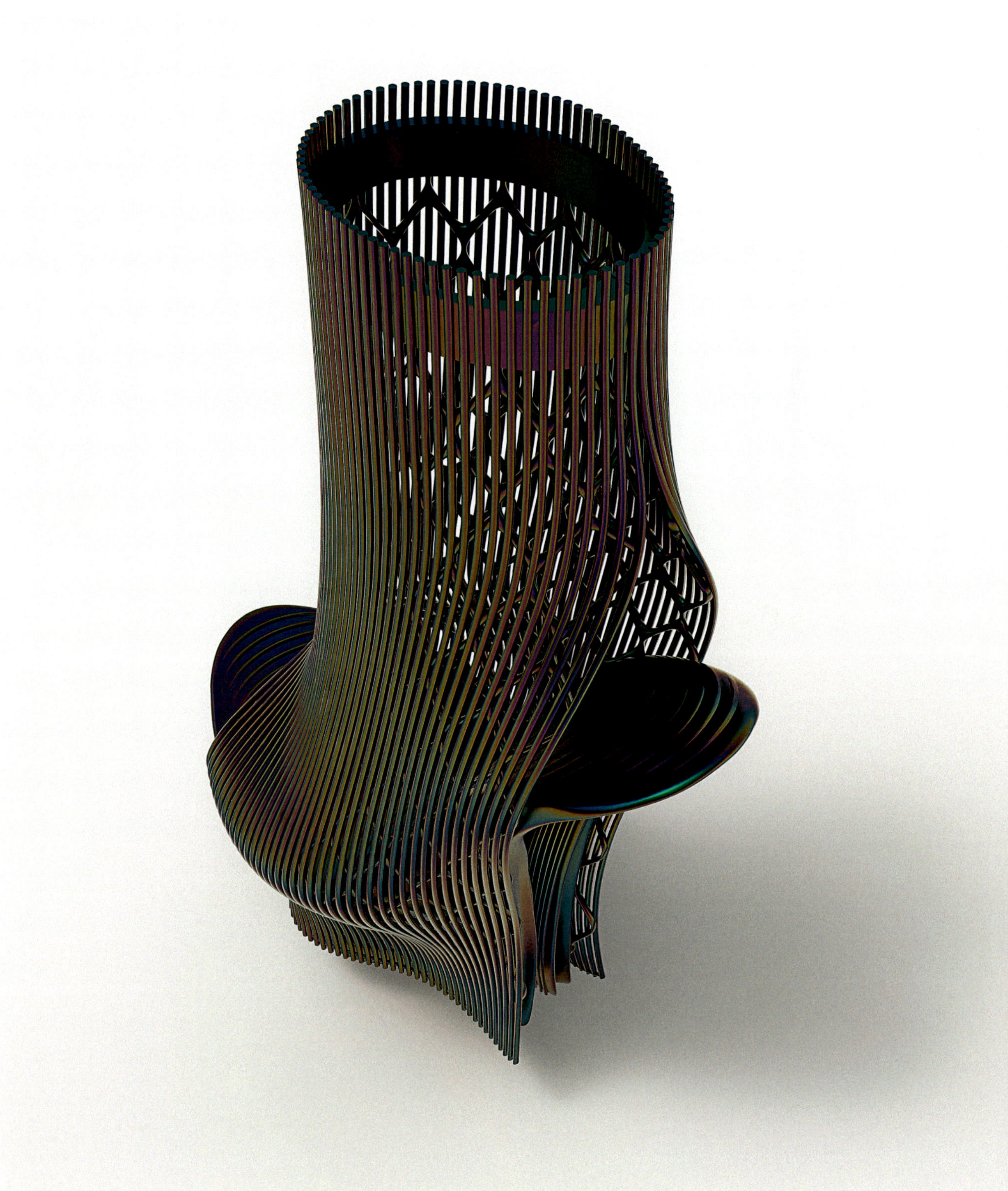

Ilabo shoe, 2015

See pp. 106–107

Digital

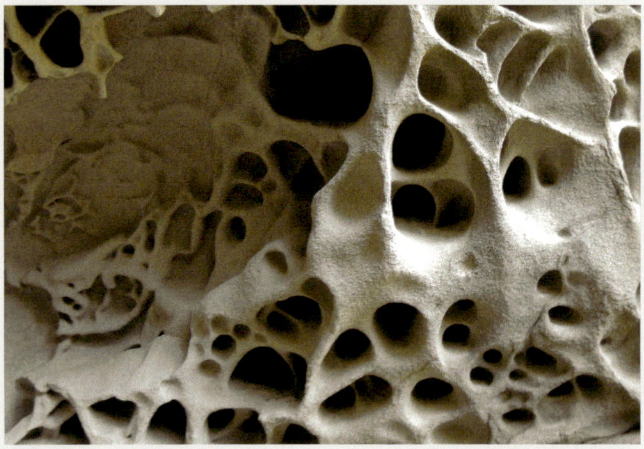
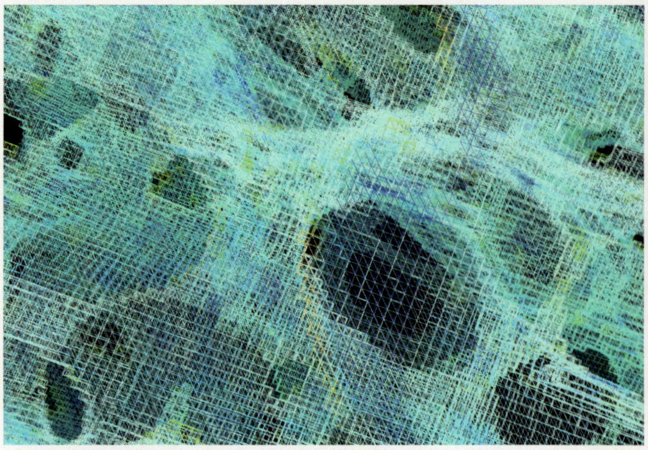
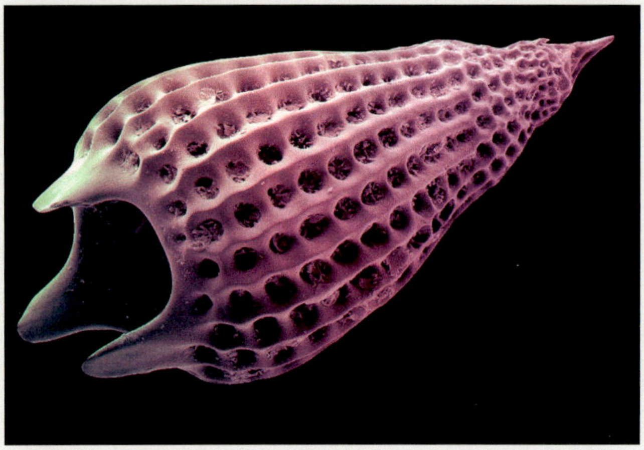
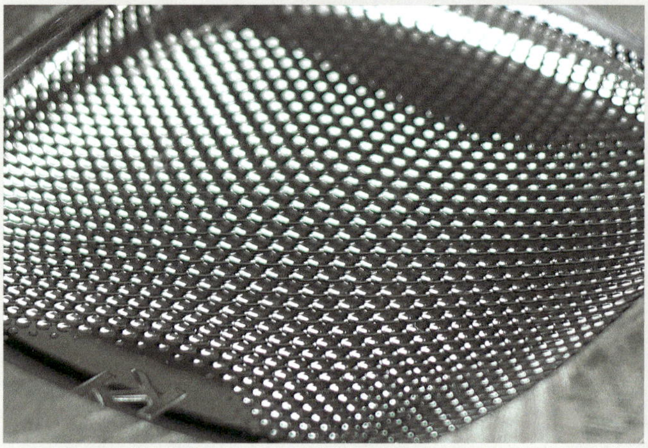
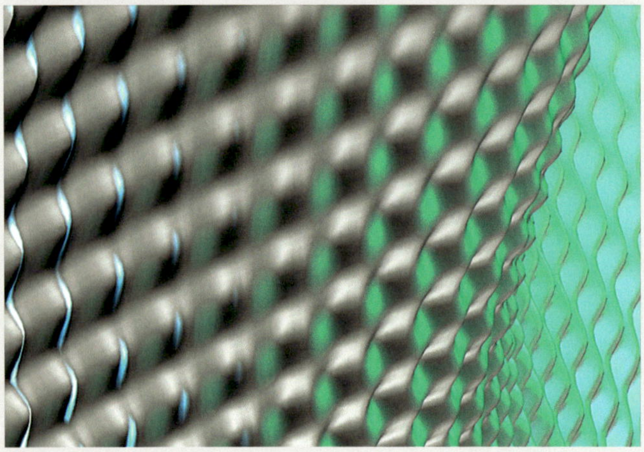
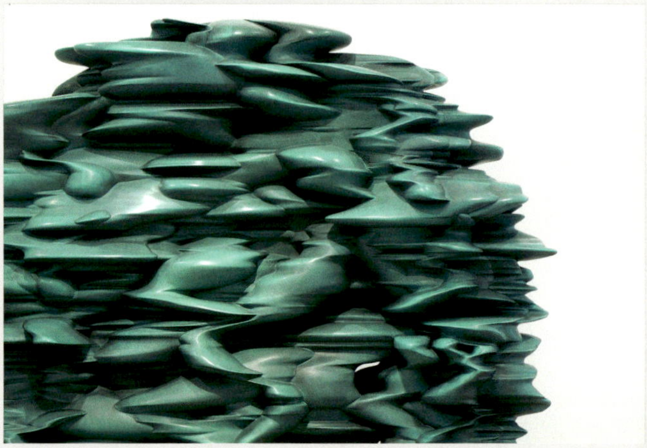

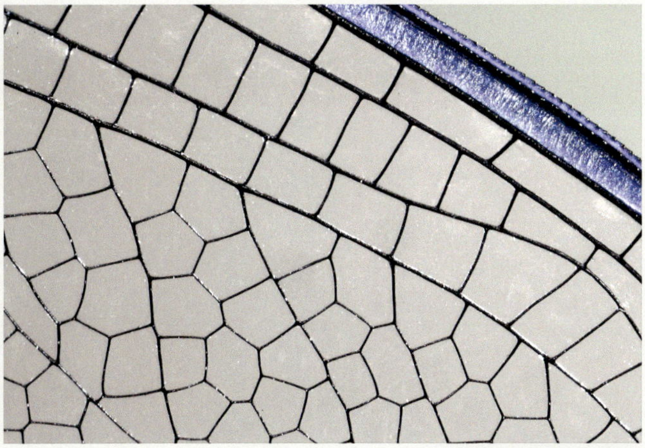
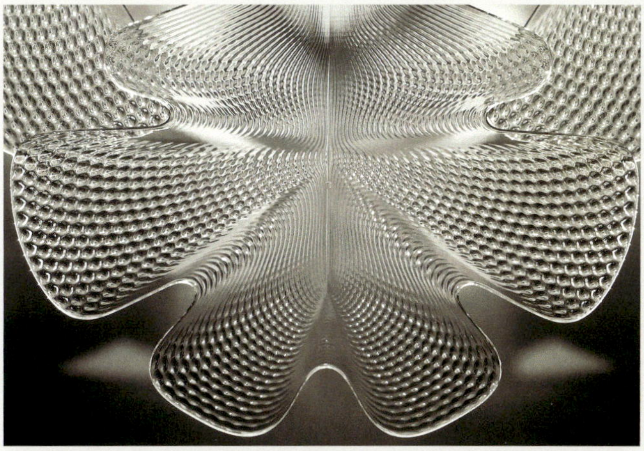
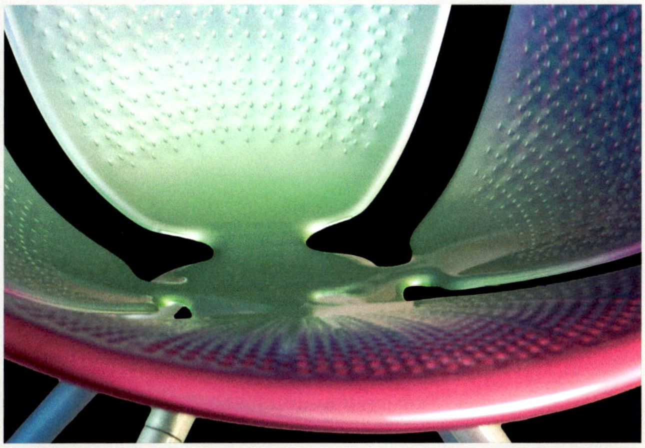
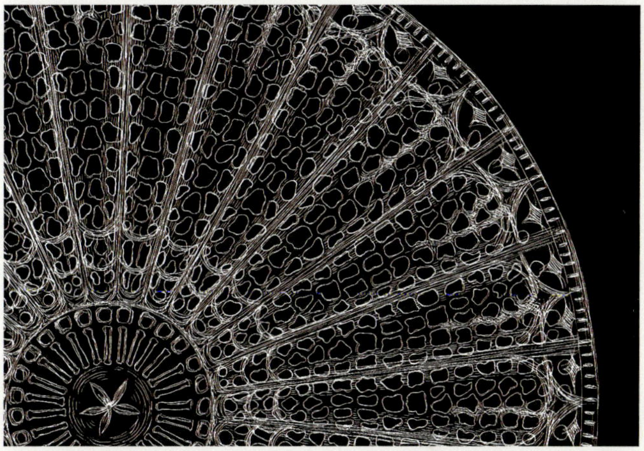
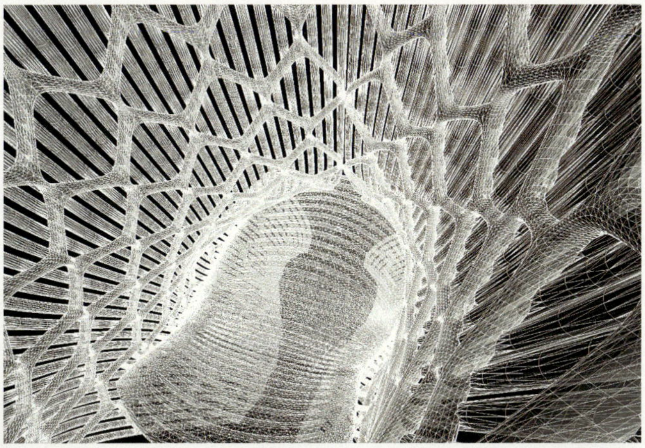
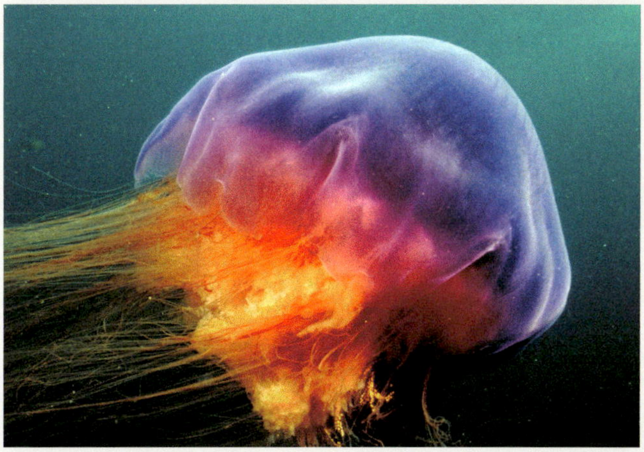

Page 98 *(from left to right and top to bottom)*

Sandstone in which the cavities have been carved out by rainwater, forming a stone lacework, San Francisco area, United States

Ross Lovegrove, *Cellular Automata* (detail), 2008, computer-generated image for this group of 5 printed studies in stereolithograph, Ross Lovegrove Collection

Radiolarian frustule (*Podocyrtis mitra*), seen through a scanning electron microscope [colored by the Ross Lovegrove Studio]

Ross Lovegrove, injection molds for the Wanderlust collection by Kiko, 2016

Ross Lovegrove, Cosmic Leaf lamp (detail), 2009, steel and chrome, polymer; H.: 192 cm, L.: 38 cm, Dia. (base): 22 cm, manufacturer: Artemide, Italy

Tony Cragg, *Versus* (detail), 2011, bronze, 150 × 160 × 55 cm, no. 13572, Marian Goodman Gallery

Page 99 *(from left to right and top to bottom)*

Dragonfly wing (*Anax imperator*), Netherlands [close-up, partially colored by the Ross Lovegrove Studio]

Ross Lovegrove, Chlorophilia 2 suspension light (detail), 2015, polycarbonate and aluminum; H. max.: 200 cm, Dia.: 77.5 cm, manufacturer: Artemide, Italy

Ross Lovegrove, Diatom chair (detail), 2014, natural aluminum molded with colored lacquer; 77 × 72 × 63 cm, manufacturer: Moroso, Italy

Christina Brodie, drawing of a central diatom frustule (*Arachnoidiscus ornatus*), microscopic monocellular algae

Ross Lovegrove, Ilabo shoes, computer-generated image (detail), 2015, 3-D printed nylon shoes; 35 × 30 × 12 cm, manufacturer: United Nude, Netherlands

Lion's Mane jellyfish (*Cyanea capillata*), Atlantic Ocean, Nova Scotia, Canada

At the beginning of the 1990s Ross Lovegrove was something of a pioneer, being one of the first designers to integrate digital tools into his work. As early as 1990 the Eye Digital Camera demonstrated an alliance between technology and ergonomic research with its organic and fluid form. Between 1999 and 2001 the new possibilities offered by digital technologies led to the creation of the Tŷ Nant mineral water bottle, a key project resulting from a long reflection on the nature of water. This constant fascination for purity and fluidity resulted, ten years later, in the Lasvit LiquidKristal pavilion, exhibited in Milan in 2012, in which the undulating glass panels were made possible by simulation software. Movement, data, immateriality, algorithmic processes—the work of Ross Lovegrove takes account of a dematerialization that itself takes into account the creation of forms growing according to a generative process rather than by construction. Through the study of the evolutionary principles of natural forms, the designer creates simple and functional objects whose complexity however lies in their development. Uniting innovative materials with a lightness of form and use, these objects combine an efficiency of form reduced to the essential and a minimal use of energy. Three-dimensional printing can also be used in the development of unpredictable processes, as in the *Cellular Automata* (2008), created from a scan of bone tissue from a human forearm. The three organic-shaped lights from the Cosmic series, which Lovegrove created for Artemide between 2009 and 2011, were also designed using a computer algorithm. In 2015, for the Ilabo shoes, he started with the data from a medical scan in order to create the bionic geometry of a shoe perfectly matched to the morphology of a foot. The result, printed in 3-D, is an amalgamation of design, architecture, and bioengineering.
As Lovegrove sees it, the designer must be an interpreter of technology. Throughout his career, he has never ceased questioning technology to find possible applications and to explore how, with its cooperation, design can become art.

Olivier Zeitoun

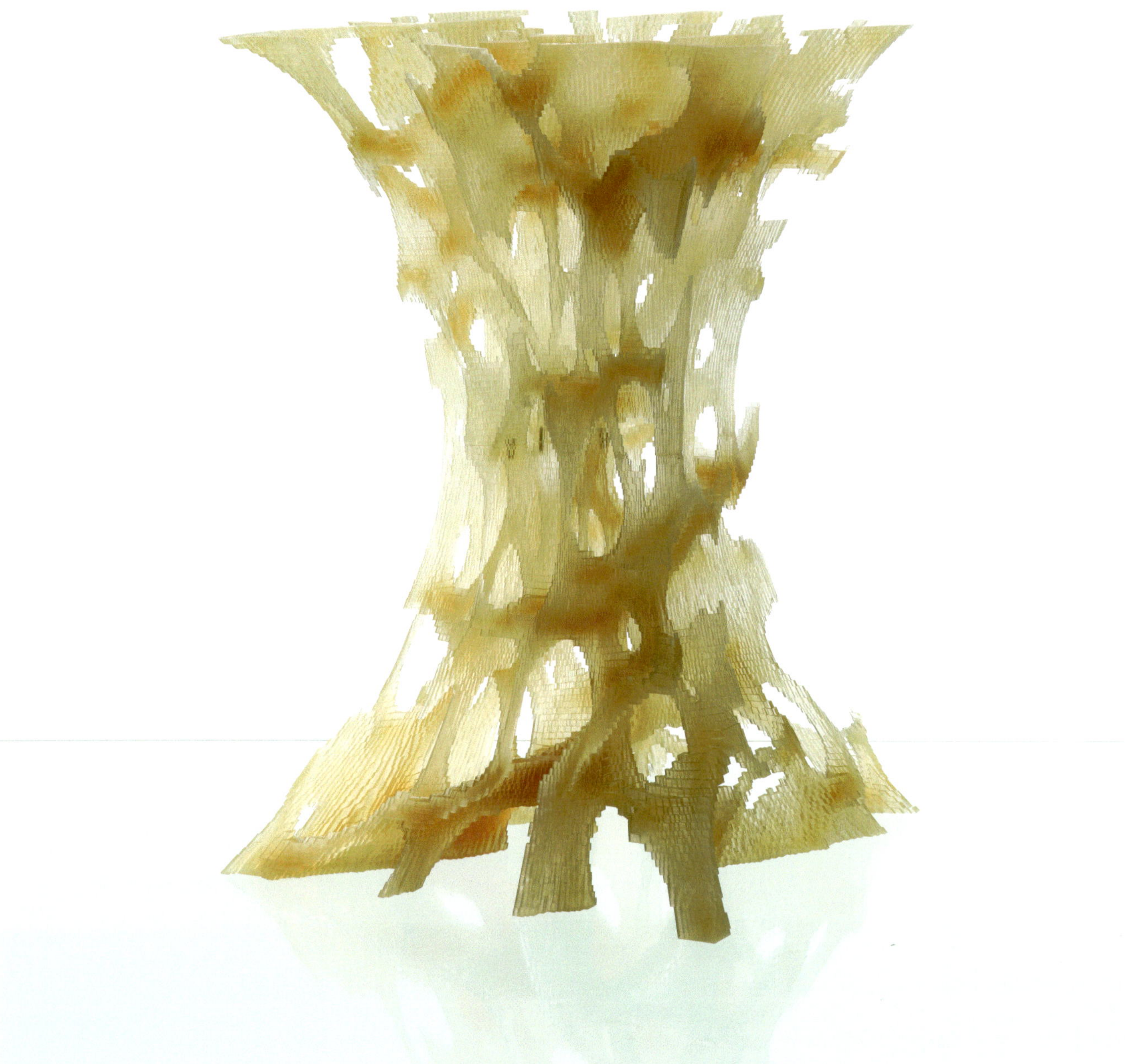

Cellular Automata

2008
Research group printed in 3-D
5 unique pieces
Opposite: stereolithograph; 60 × 45 × 45 cm
Below: computer-generated images

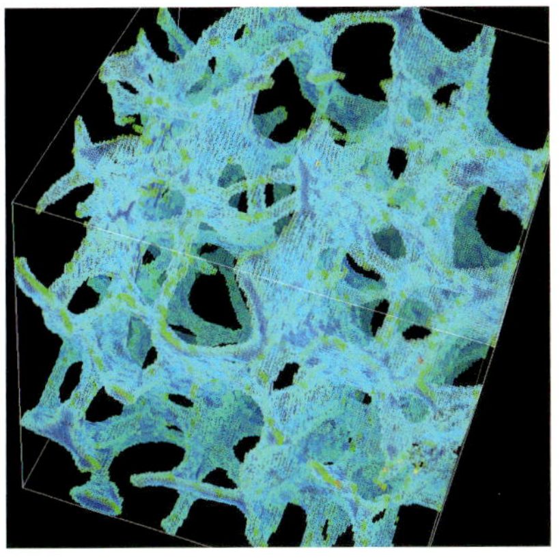

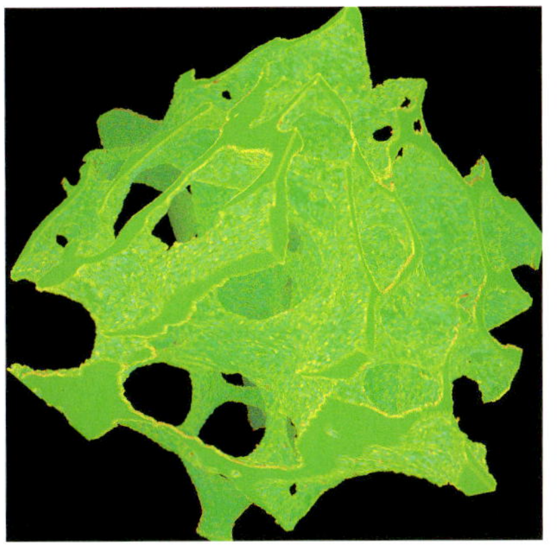

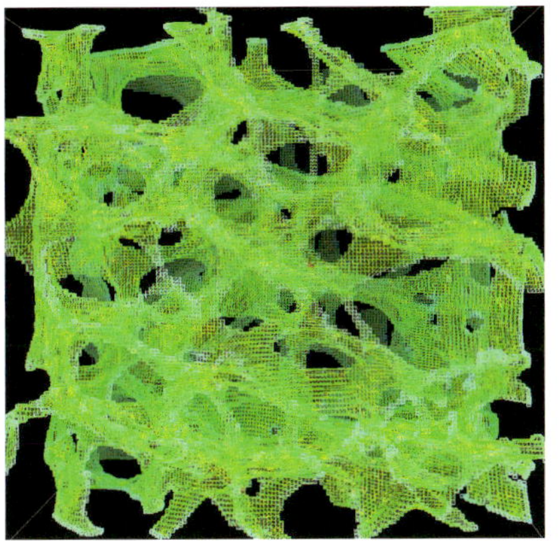

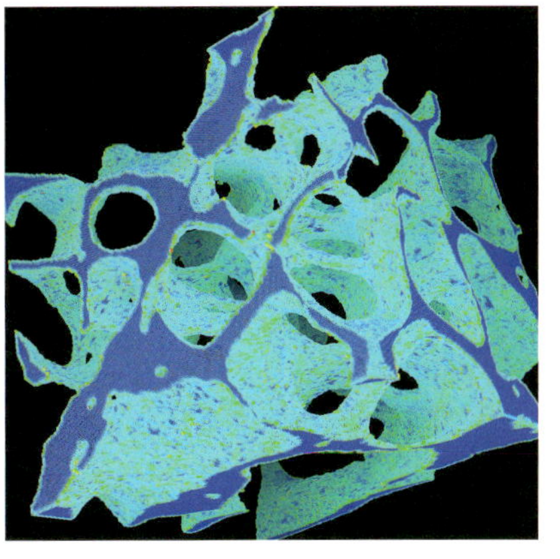

The five stereolithographic studies that comprise *Cellular Automata* were made from a bone scan of a human forearm. Each of these studies was enlarged to make the living bone tissue and its cellular structure visible to the naked eye. The analysis consisted of observing their composition and structure, paying particular attention to the Wiener process—a mathematical description of Brownian motion, which a fluid undergoes when a large particle is immersed in it—in order to apply this phenomenon to other fields, such as design. The resulting product is a three-dimensional form that is both strong and light. *Cellular Automata* also utilizes algorithms designed from cellular automatons used in parametric architecture.

MV

Nebula wall lighting piece

2012
Aluminum, opaline methacrylate, polished chromed metal, LED; Dia.: 80 cm, D.: 14 cm
Manufacturer: Artemide (Italy)

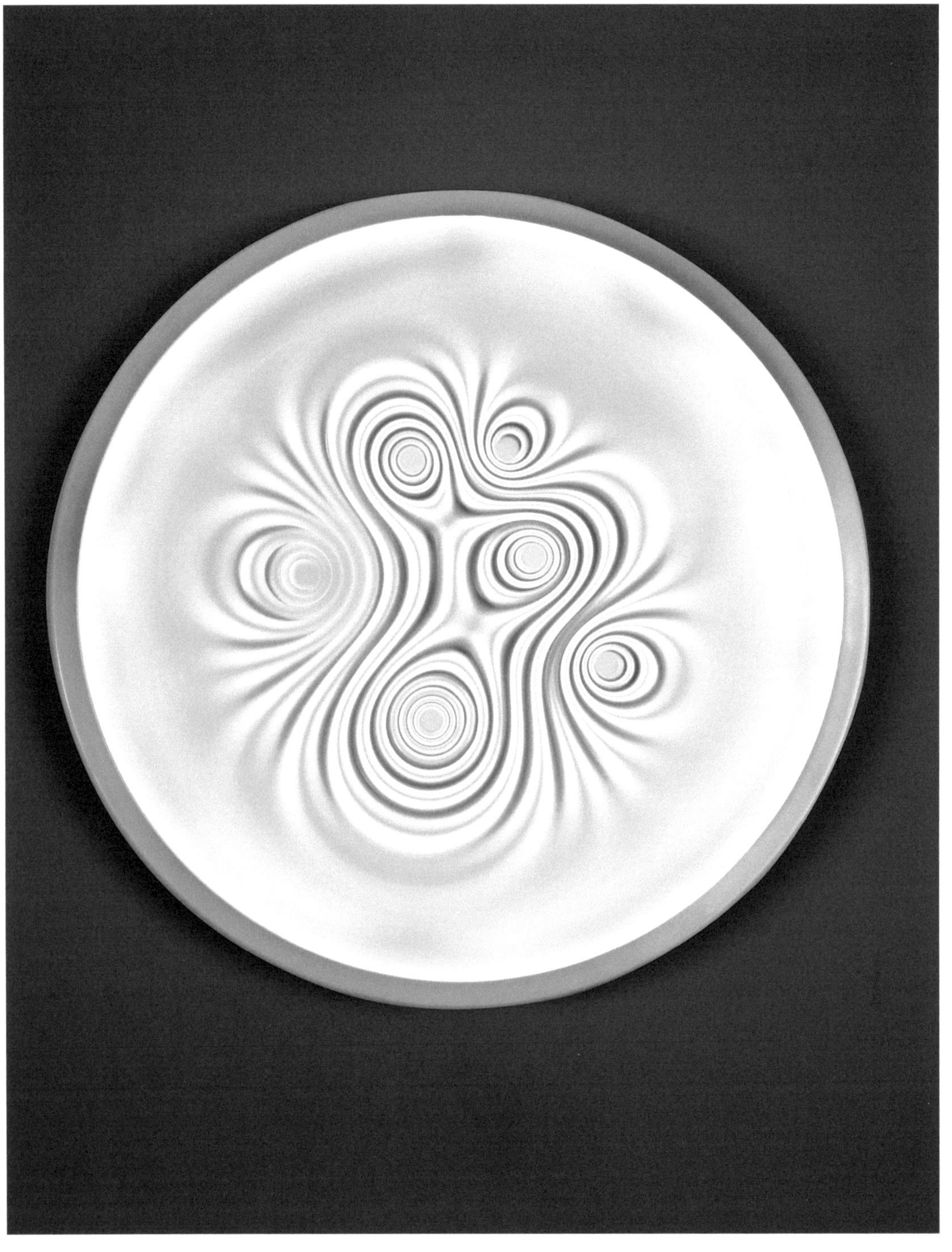

Net Architecture

2005–2006
Architectural project model:
3-D printed:
H.: 140 cm, Dia.: 28 cm

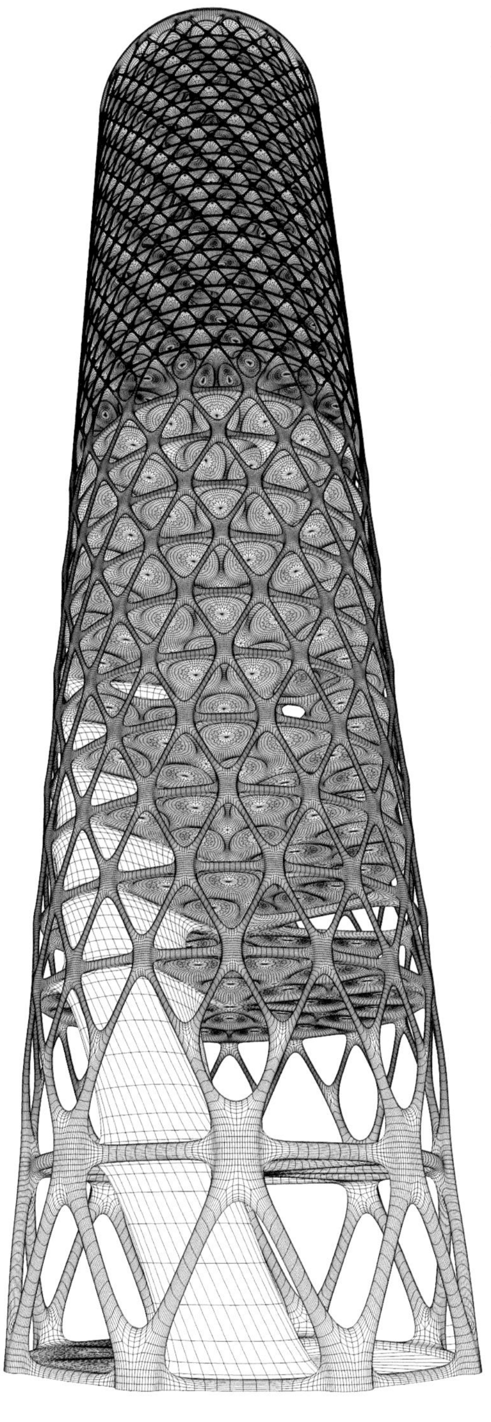

This "net architecture" project followed a study on the evolution of 3-D printing, which highlighted the generative density and geometry applicable to a polyfunctional architecture situated in an urban area. The anticipatory structure that resulted from this research was inspired by Jean Nouvel's uncompleted project *Tours sans fins* (1988). Lovegrove's tower is printed in carbonized concrete, a material that ensures the structural integrity of the object. The tower is made up of three levels: a ground floor composed of high-vaulted units used for small shops; a first floor used as office space; and upper floors intended for housing. Each of these levels is oriented in a direction that optimizes the circulation of natural light. An asymmetrical elevator creates the central point and the spinal cord linking all the levels of the structure.

MV

Ilabo shoes

2015
3-D printed nylon and polyurethane; 35 × 30 × 12 cm
Edition of 50 and 1 artist's proof
Manufacturer: United Nude (Netherlands)

The Ilabo shoe, limited to an edition of fifty pairs, is the result of the "Re-inventing Shoes" project launched in 2015 by the United Nude company in partnership with 3-D Systems. Five renowned architects and designers (Ben van Berkel/UNStudio, Zaha Hadid, Fernando Romero, Michael Young, and Ross Lovegrove) took part in the project and each produced a model for a woman's shoe, destined to be entirely printed using the selective laser sintering (SLS) process. The participants were given one precise specification: to use a rigid material (polyamide) for the sole and another, flexible material (polyurethane) to provide the shoe's comfort.
Lovegrove's pair of Ilabo shoes was first modeled, using a human body, by the engineer Arturo Tedeschi, an expert in the computer modeling program Grasshopper 3D. Put more precisely, Tedeschi and Lovegrove used the data from a 3-D medical scanner to bring together the attraction/repulsion logic in the bionic geometry of the foot, and thus to obtain a shortened form that gives an impression of verticality, while at the same time eliminating the bulk and weight of the shoe. The materials used, nylon and thermoplastic polyurethane colored prior to printing, formed a geometric, three-dimensional mesh that is both fluid and organic. The use of advanced 3-D printing methods enabled the realization of a free, unfettered architecture that supports the anatomy of the women's foot, which itself is left exposed at the heel and toes.
This convergence principle, fundamental in Lovegrove's work, is present here in the form and the structure, which are created using studies of foot morphology and the foot itself, once it is encased and emerging from the shoe. The Ilabo shoes are thus positioned at the crossroad of design, architecture, and bioengineering.

MV

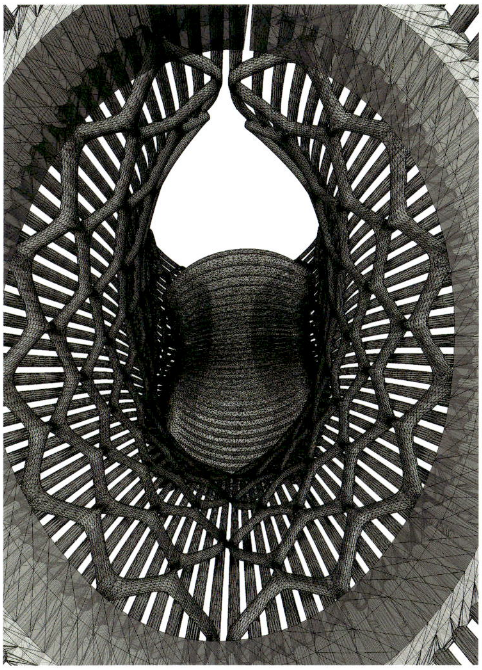

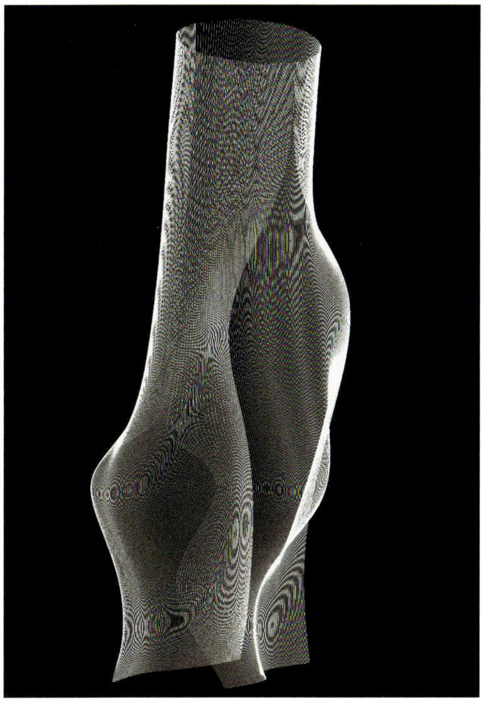

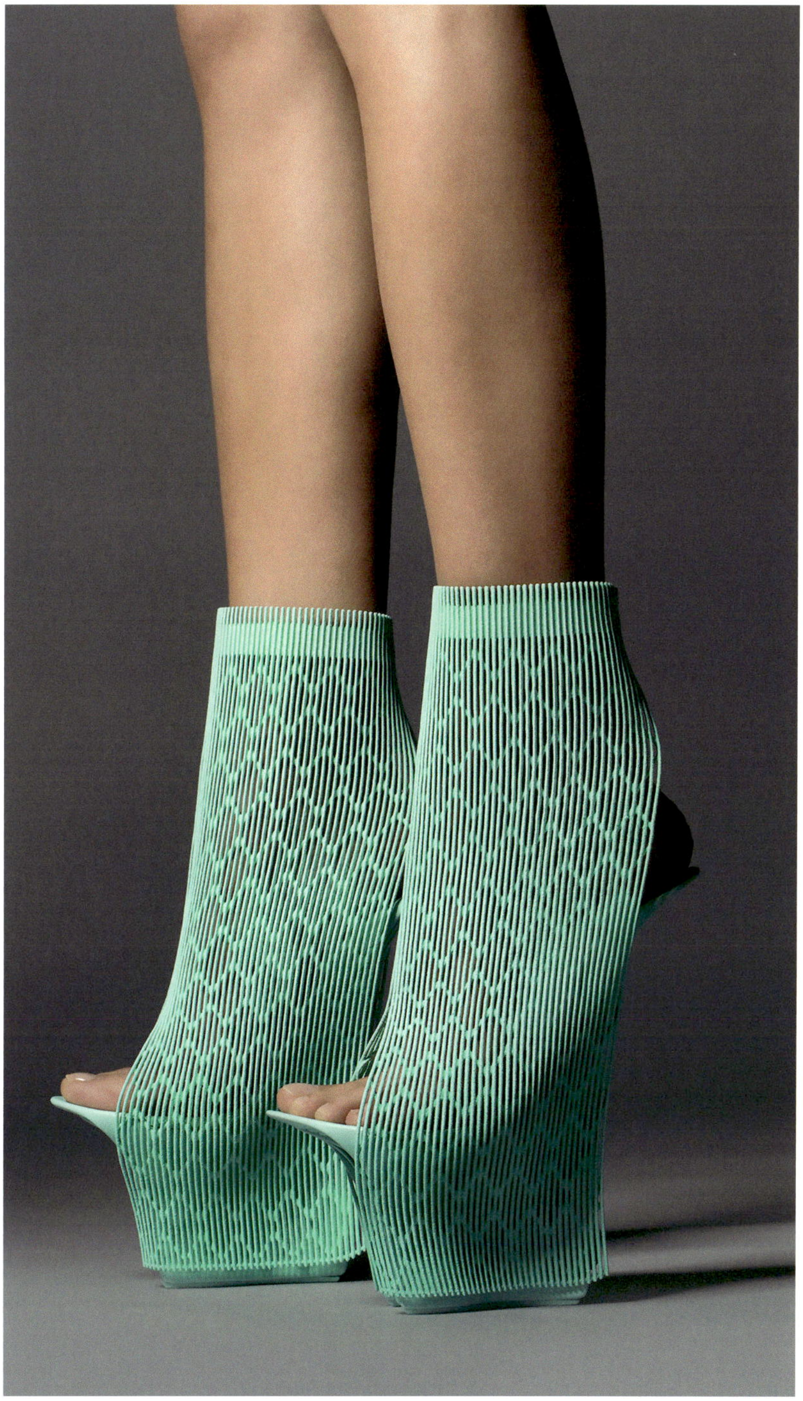

Cosmic Angel light

2009–2011
Ceiling light
Aluminum, polymer; 25 × 178 × 80 cm
Manufacturer: Artemide (Italy)

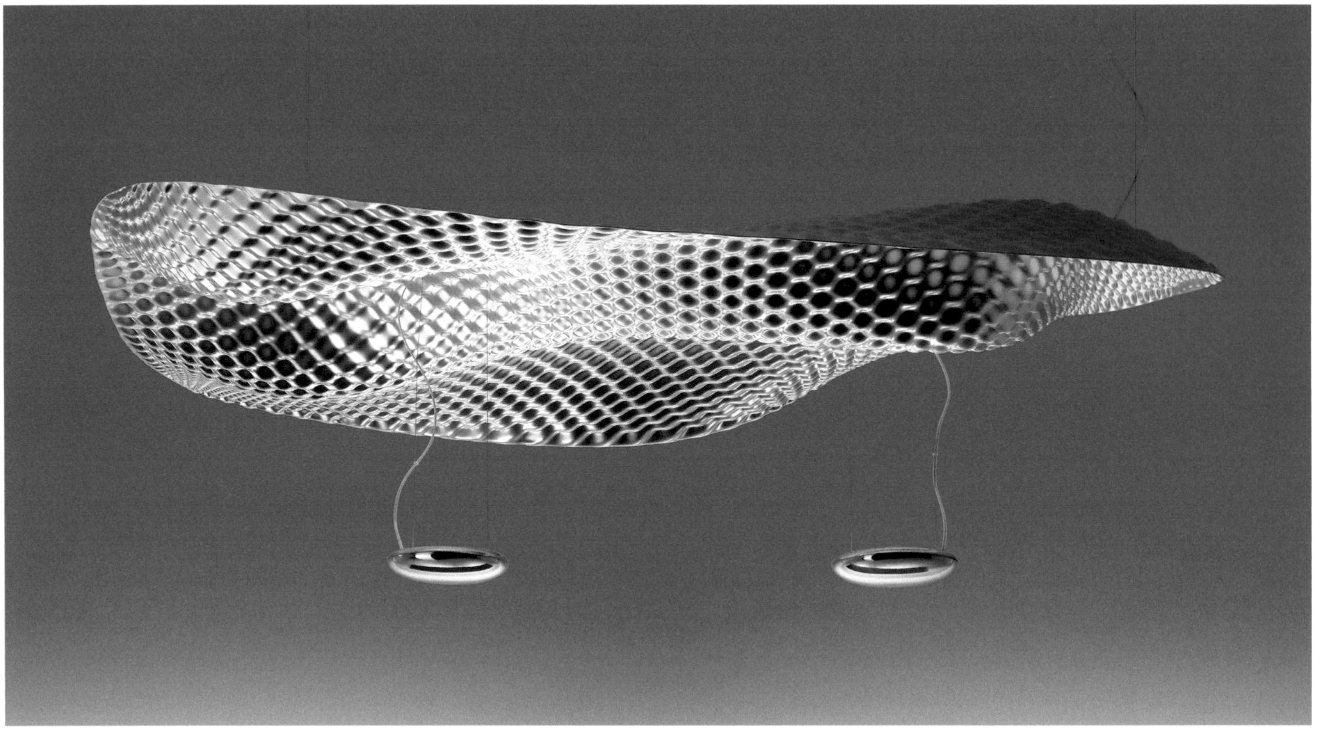

Made for the Italian firm Artemide, Cosmic is a series of lights created between 2009 and 2011 that includes Cosmic Angel, Cosmic Leaf, and Cosmic Landscape. This series circumvents physicality in favor of an ambiance of light. At once simple and complex, the fluid, organic forms are created using algorithmic calculations. Each of the lights is distinctive. Cosmic Angel is produced from polymer (ABS) molded in a vacuum, a technique inspired by a Greg Lynn project for Alessi. The ceiling light has a multisensorial surface, made up of an undulating aluminum structure, assembled to make a metallic mesh resembling scales. Through its design, the surface captures light from seven different sources, which transforms its color as in a dichromatic system. The diffracted light is thus given an ethereal quality.

MV

Chlorophilia 2 suspension light

2015
Ceiling light
Polycarbonate, aluminum, LED;
H. max.: 200 cm, Dia.: 77.5 cm
Manufacturer: Artemide (Italy)

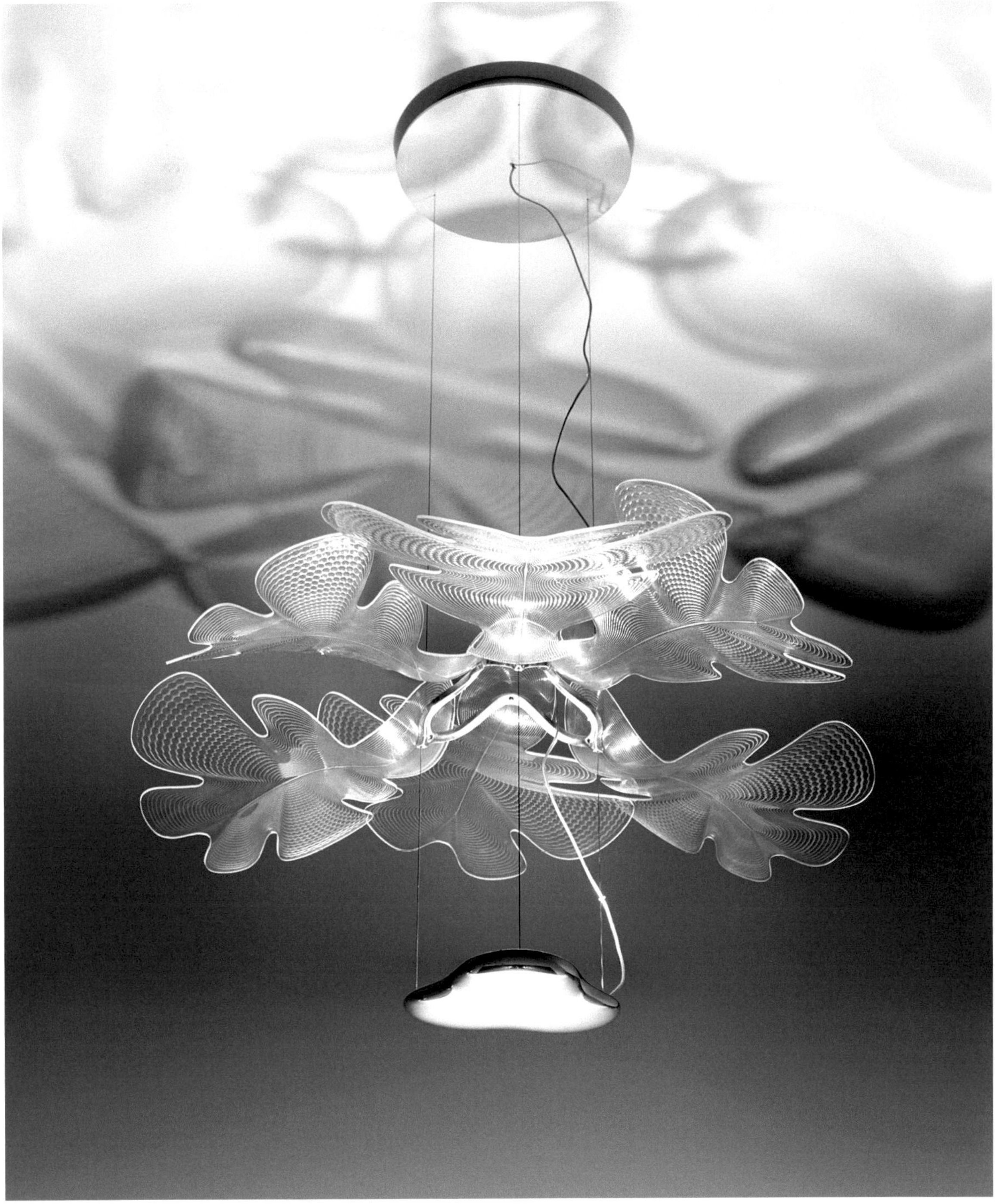

Diatom chair

2014
Natural molded aluminum varnished with a glossy transparent protective coating; 77 × 72 × 63 cm
Manufacturer: Moroso (Italy)

For this project designed for the Italian firm Moroso, Ross Lovegrove was inspired by microorganic, unicellular forms from the seabed, in particular the frustule (external skeleton) of the planktonic algae called diatom. In order to make use of the workings of this radiolarian, he studied the mathematical rules governing its vegetative growth. Lovegrove then devised applications that best corresponded to the logic of industrial production in terms of structure and material.
Stackable and very light, the Diatom chair was designed entirely by computer and made in molded aluminum. The parametric construction methods, also used in digital architecture, allowed all the elements of the structure to be made malleable by parametrics, and all the systems between them to be linked to obtain a complex form. The seat of the Diatom chair was developed by a procedure of hydroformed compression, in which a sheet of aluminum molded under pressure is shaped into a three-dimensional form. The feet of the chair are soldered to the bowl-shaped seat to produce tubular supports. The two techniques are thus brought together to make one complete piece. Having compressed the aluminum surface with the help of a rationalized series of micropoints, reinforced in specific places, the chair is much lighter than if it had been made with steel. Moreover, its reduced bulk and the structural density of its cross-linked surface, obtained using techniques close to automobile and aeronautical engineering, reinforce the properties of the material.
MV

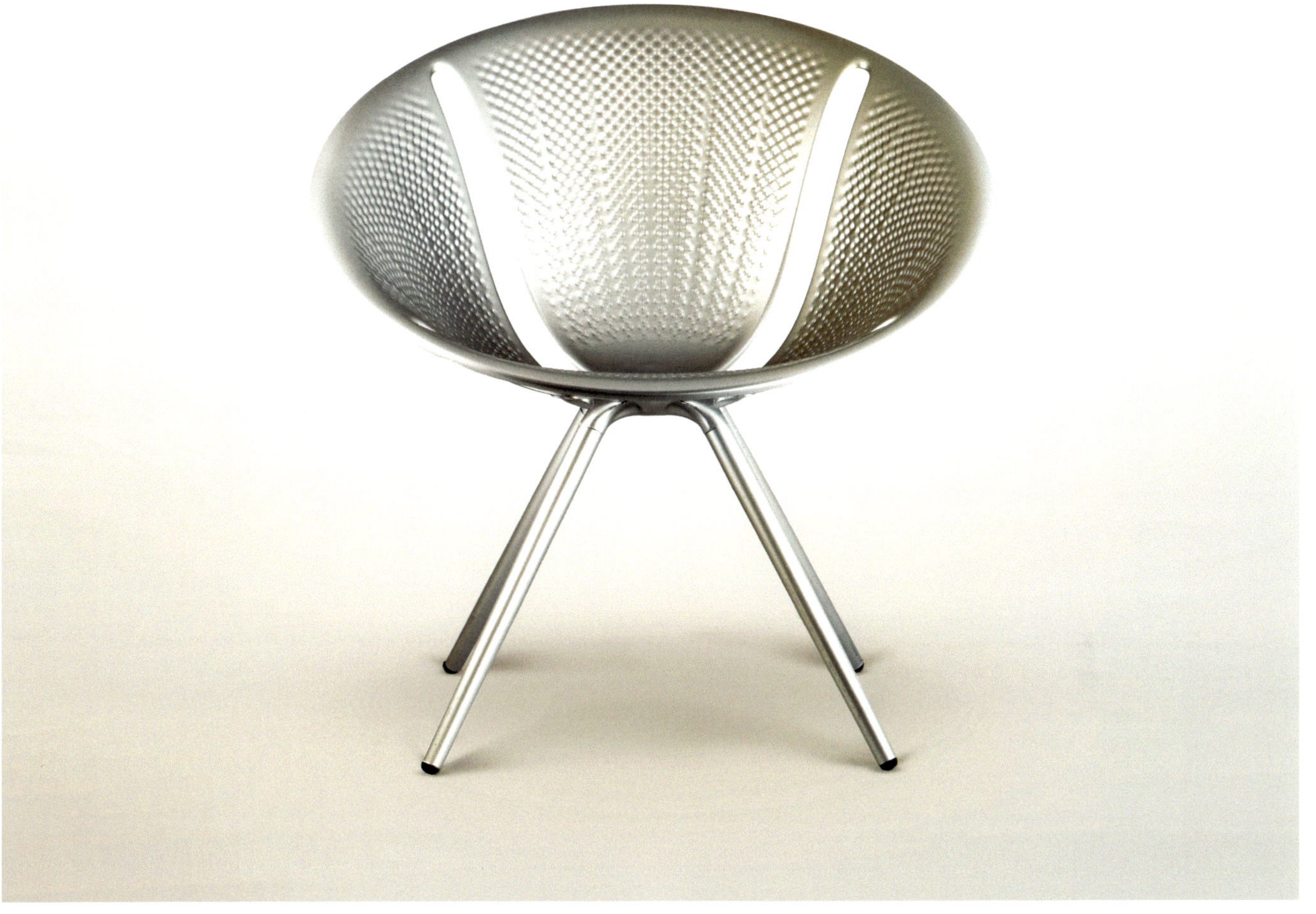

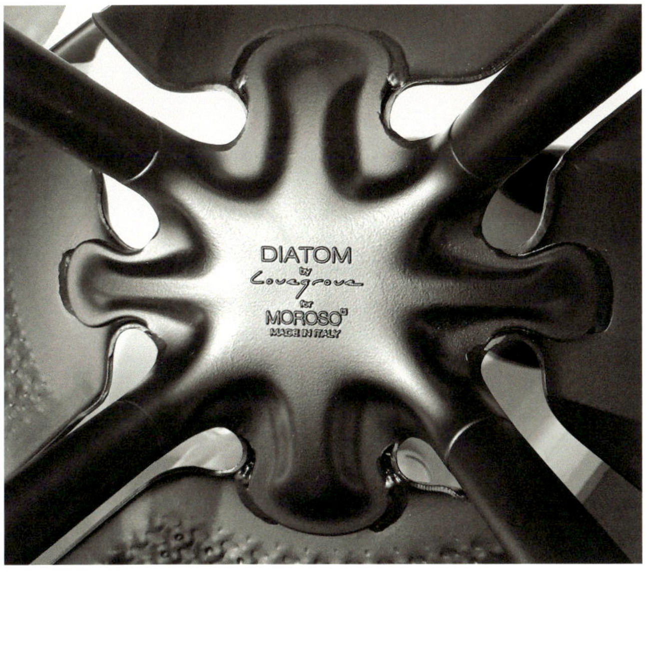
DIATOM
by
Lovegrove
for
MOROSO
MADE IN ITALY

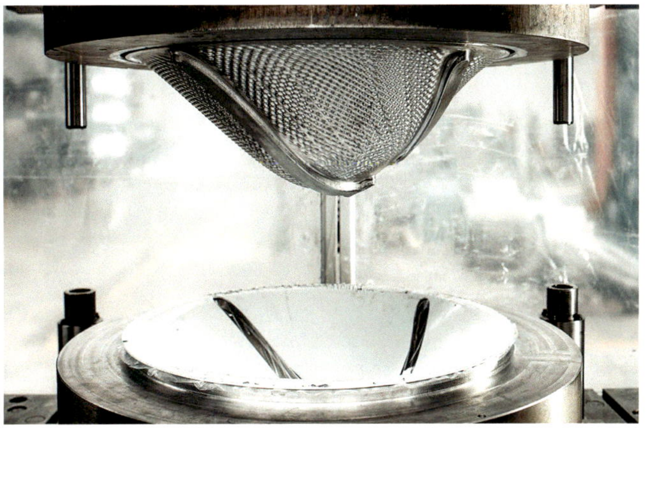

Corolised Chair

2012
Alumide printed in 3-D; 70 × 45 × 45 cm
Edition of 100

"Corolization" is a term used to define the growth process of structures rich in calcium, like coral. This process corresponds more precisely to the creation of a voluminous mass, which filters and protects several life forms. It is caused by the deposit and subsequent calcification of marine microorganisms. During this process, governed by the ebb and flow of sea currents, the microorganisms take on complex biomorphic forms. From this point of view, corals are particularly interesting to study. They are a very rich source of decorative applications. For example, the *Diploria labyrinthiformis,* better known by the name of "brain coral," is remarkable for the multiplicity of its convolutions, akin to that of a human brain. The study of the evolution of the structures and their relationship to the workings of the human organism have been at the center of Lovegrove's work: "I make design objects as if they were distinct pieces. However inanimate they may be, they fit in with a structural, material, and adaptive origin, which converges with the atmospheric conditions, gravitational forces, the context, and the use to which it is put." Applied to Lovegrove's work, corolization fits in with the research into the dematerialization of the mass of products. The designer came up with the idea for this project during a visit to the Judd Foundation in Marfa (Texas), where he was able to observe more than fifty varieties of a single cubic structure. The three pieces on view are part of a production program for one hundred chairs printed in 3-D at a scale of 1:1, based on an organic geometry reminiscent of the work of Carlo Mollino.
MV

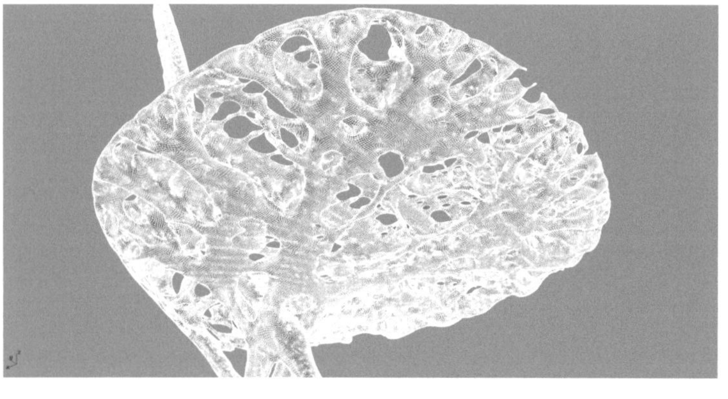

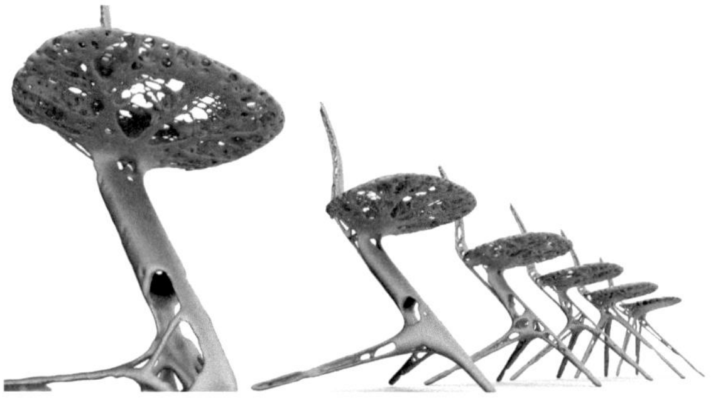

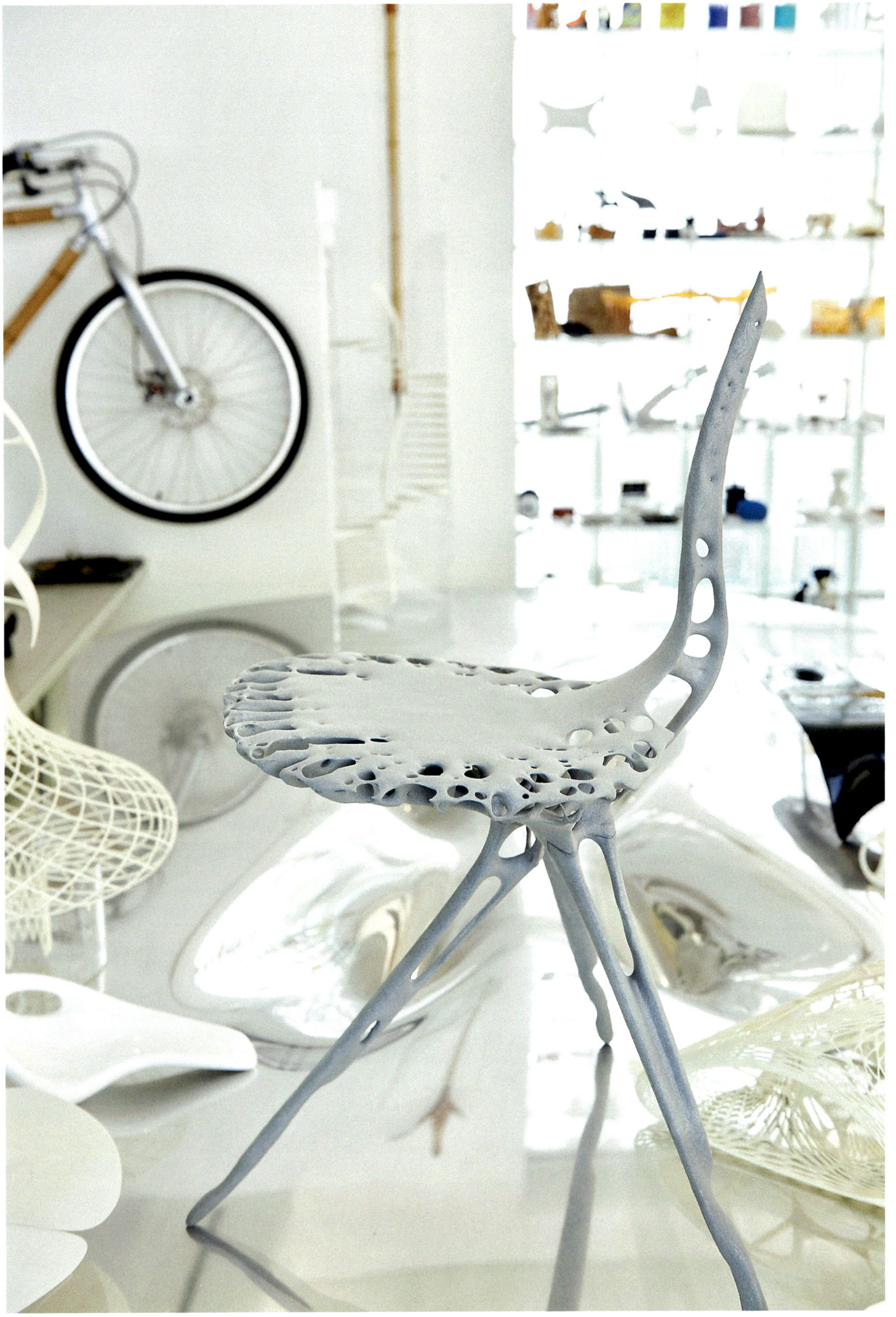

Spine Chair

2017
Prototype
3-D Print; 70 × 45 × 45 cm
Manufacturer: Moroso (Italy)

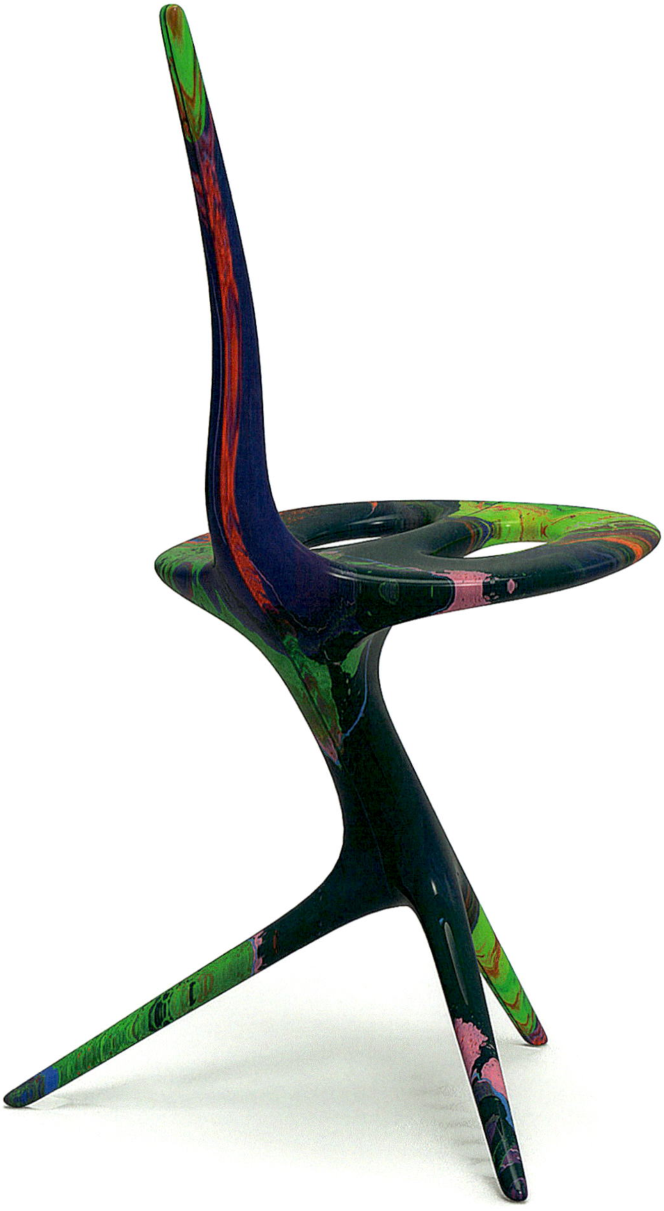

Biophilia chair

2014
Rotomolded polyethylene; 103 × 47 × 56 cm
Manufacturer: Vondom (Spain)

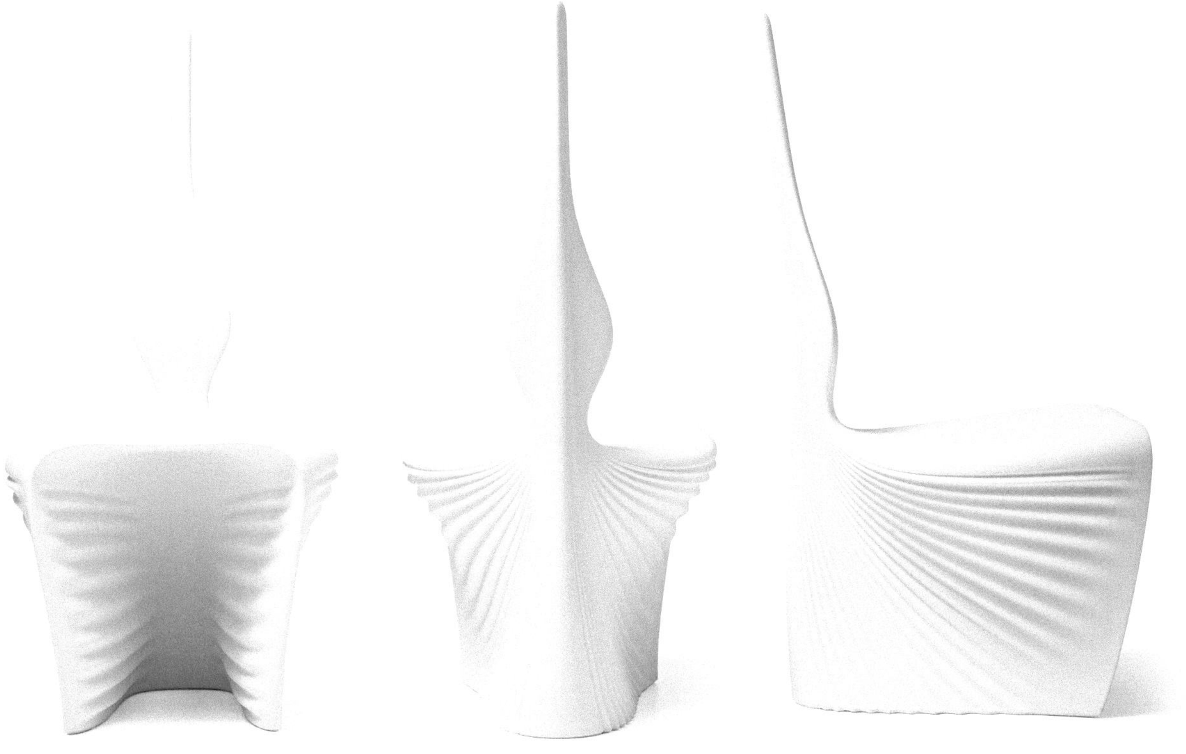

Kenzo perfume bottle

2010–2011
Opposite: drawing by Ross Lovegrove and glass prototype; 13 × 9 × 3 cm
Below: modelization studies
Manufacturer: Kenzo (Japan)

In 2010 the Kenzo brand for upmarket products commissioned a model for a men's perfume bottle from Ross Lovegrove. For this project, the designer studied the techniques for cutting flint, from which he developed a digital application. This allowed him to form a model, with the help of new materials and advanced manufacturing techniques, in the form of the "primitive" geometric shape that he sketched instinctively, inspired by these prehistoric flint tools. The result is a bottle whose abstract contours combine with its anthropomorphic aspect. This effect is underscored by the use of matt aluminum and glass, which seem to become merged into the mass of the object.

MV

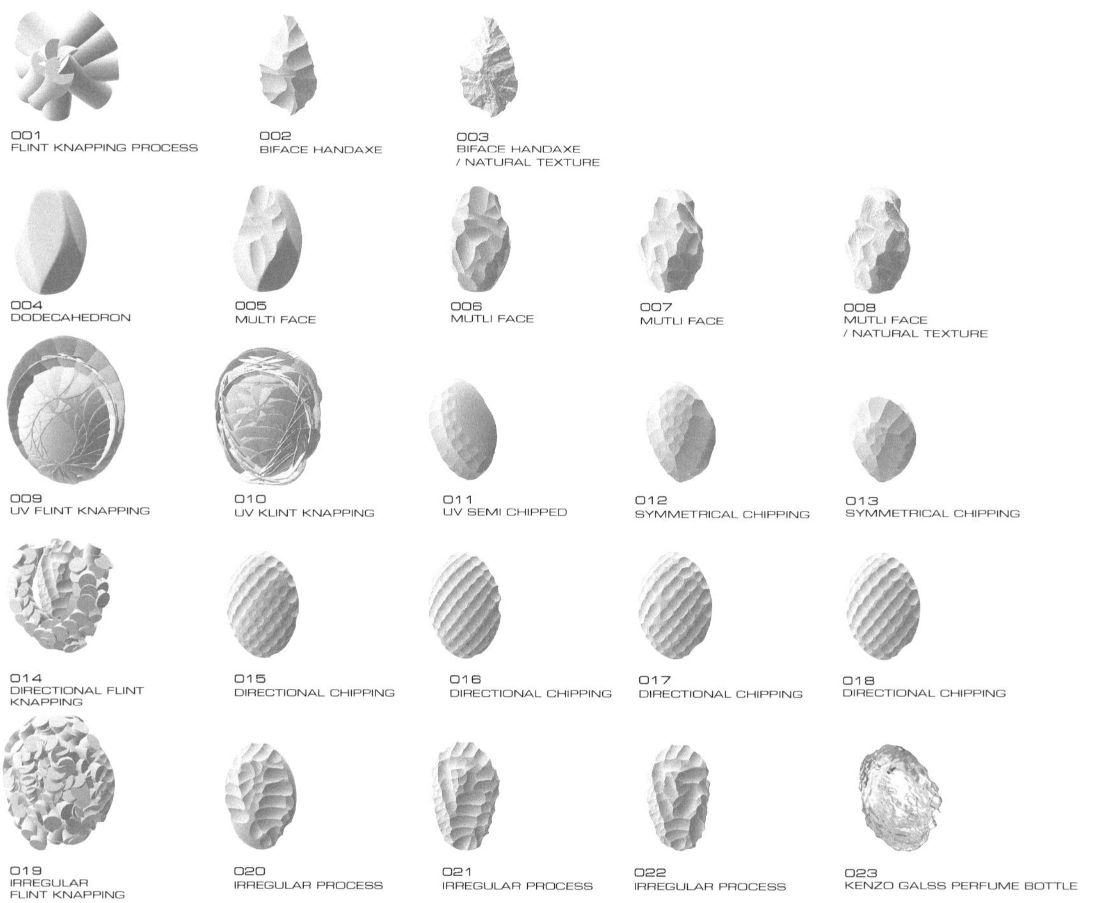

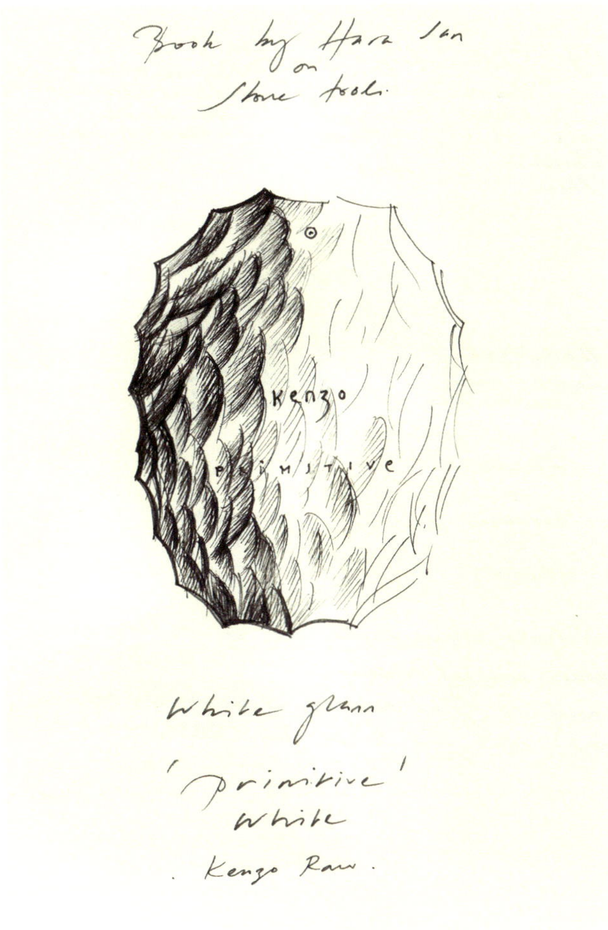
Book by Hara San on Stone tools
KENZO
PRIMITIVE
White glass
'primitive' white
. Kenzo Raw.

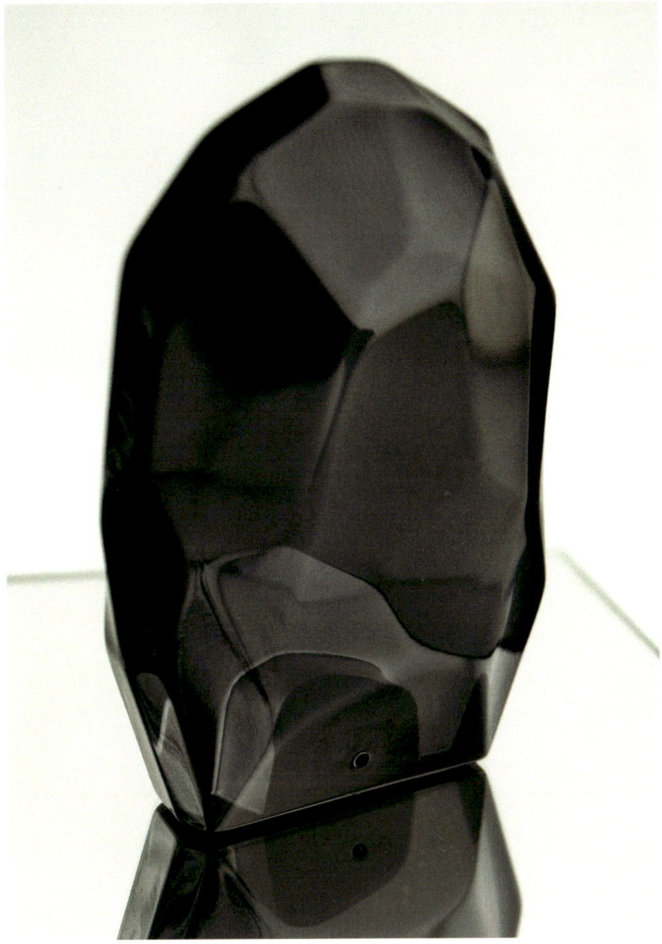

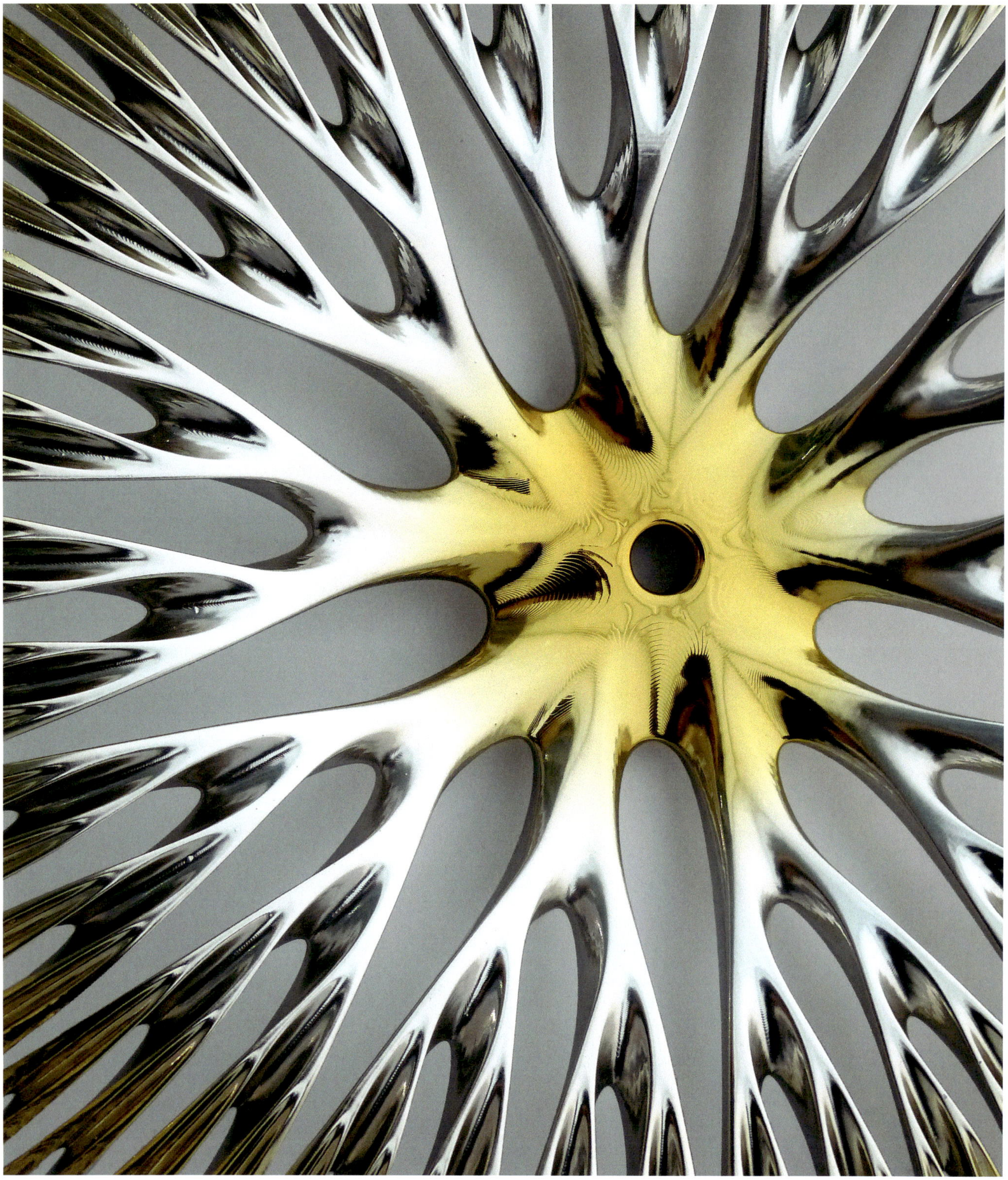

Development model for the wheels of the Twin'Z electric car, 2012–2013 (detail)

See pp. 130–139

Case Studies

Swarovski Crystal Aerospace

2005–2006
Prototype of solar-powered car
Polycarbonate, aluminum, photovoltaic cells, crystal, silicone; 100 × 420 × 180 cm
Manufacturer: Swarovski UK Ltd (United Kingdom)

With the participation of:
Sharp Solar Europe, Swarovski Optical Laboratories, Anthony Lo, artistic director for General Motors Europe, Coggiola, Turin

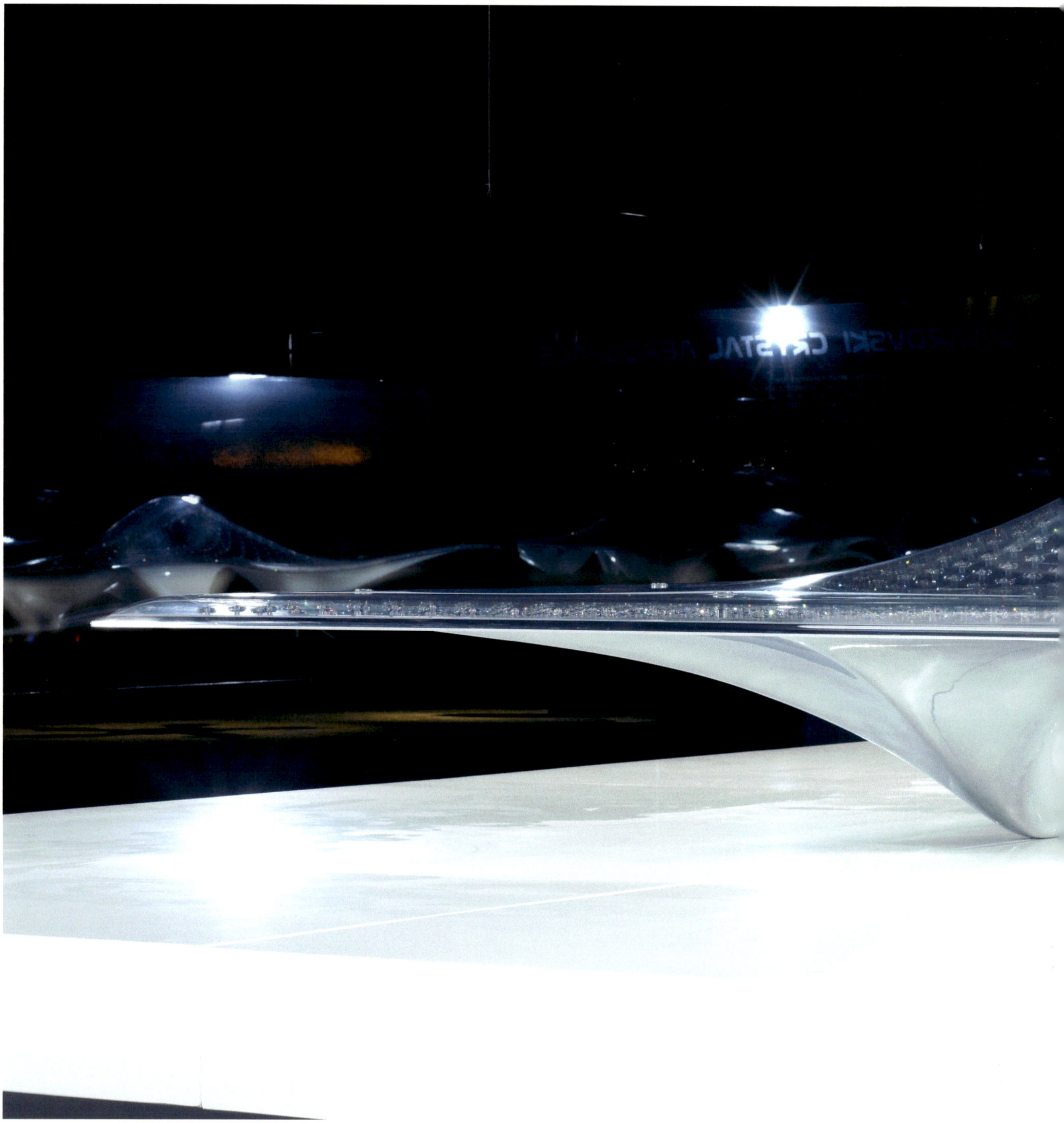

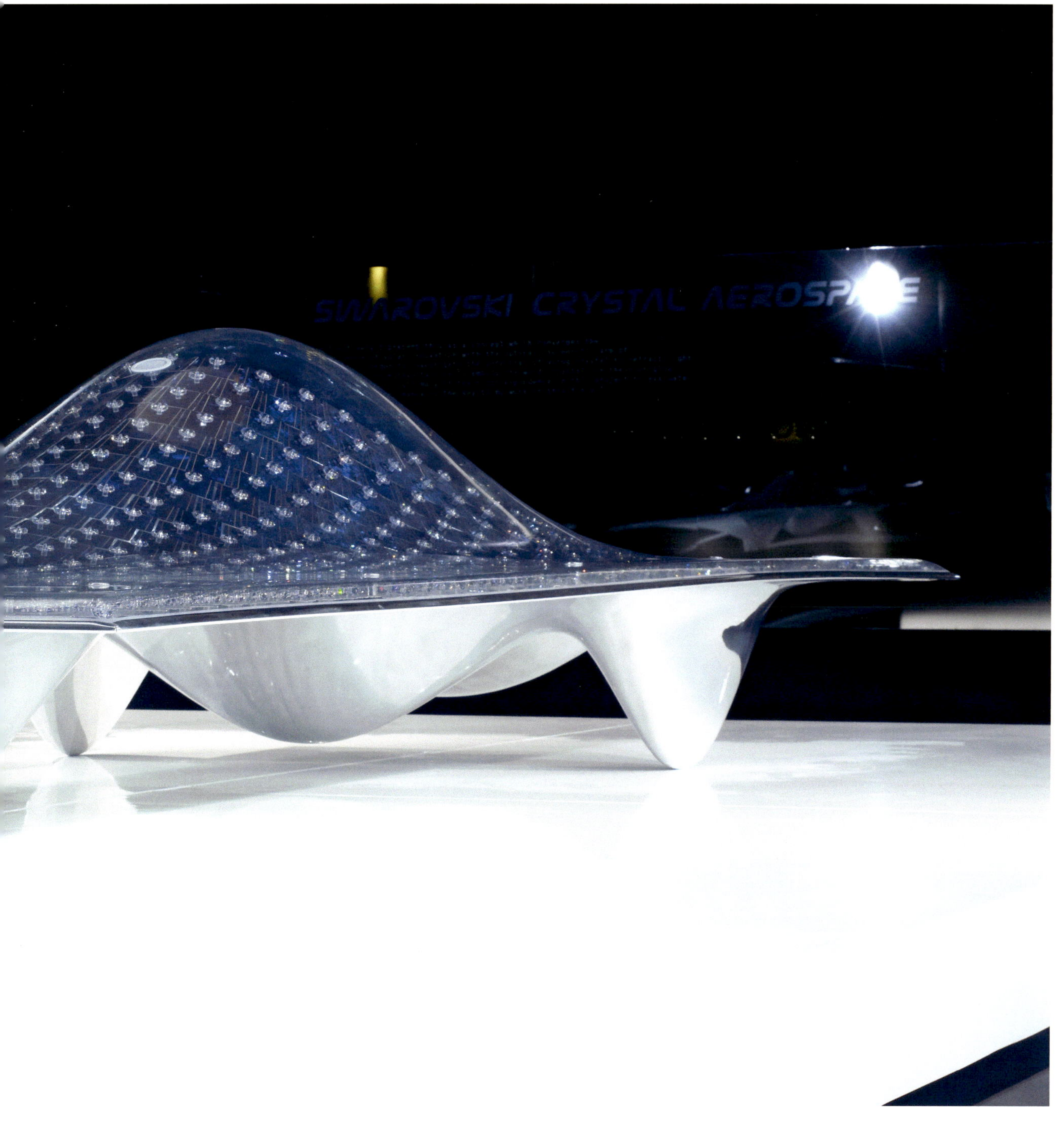
SWAROVSKI CRYSTAL

Where Art, Design, and Science Converge

Ross Lovegrove

My work with Swarovski has focused on promoting innovation by speculating on the use of crystal in unexpected ways.
Swarovski Crystal Aerospace (2005–2006) and Generator House (2006–2008) are projects linked with two major areas of universal human need: first, how we can harness the use of solar energy to achieve perpetual motion in car design; and second, how we can achieve a clean, off-grid existence via innovation in "product architecture."
Swarovski Crystal Aerospace is a polarizing and convergent example of how a single object can traverse worlds and stimulate a very new view of design in the twenty-first century. It has been exhibited extensively across the fields of art, design, technology, and transportation, giving it a unique position that is unparalleled by any other contemporary work in the last fifty years. Inspired by the intelligent engineering that permeates the space industry, the vehicle uses one thousand specially cut crystals, mounted via 3-D printed transparent supports onto one thousand polycrystaline photovoltaic panels, thus producing two states of crystal in unison. Here, sunlight is amplified and evenly distributed onto the solar panels to optimize the solar gain, using the complete upper skin to generate electricity, which is stored in lithium batteries inside the vehicle. The effect of this biomorphous blanket of cells is revelatory in its beauty born of natural materials and formulated into an aero- and hydrodynamic body form.
While creating this car, I was also assisting Cambridge University with the design of their World Solar Challenge vehicle and learned that the most expensive solar car built at that time was produced by Honda, at a cost of five million dollars. Comparing this with the average fossil-fuel equivalent car program, costing up to two billion dollars, I could see that there was an extraordinary opportunity to work with and harness the incredible integrity of such divergent modes of thinking to inspire minds from the arts and the sciences to pursue this approach.
Swarovski Crystal Aerospace upholds and promotes a new form of universal awareness toward the sort of progressive and intelligent engineering that uses materials science and physics to arrive at objects that are supremely beautiful. It has thus formed a bridge between all things considered, from art and design to science and

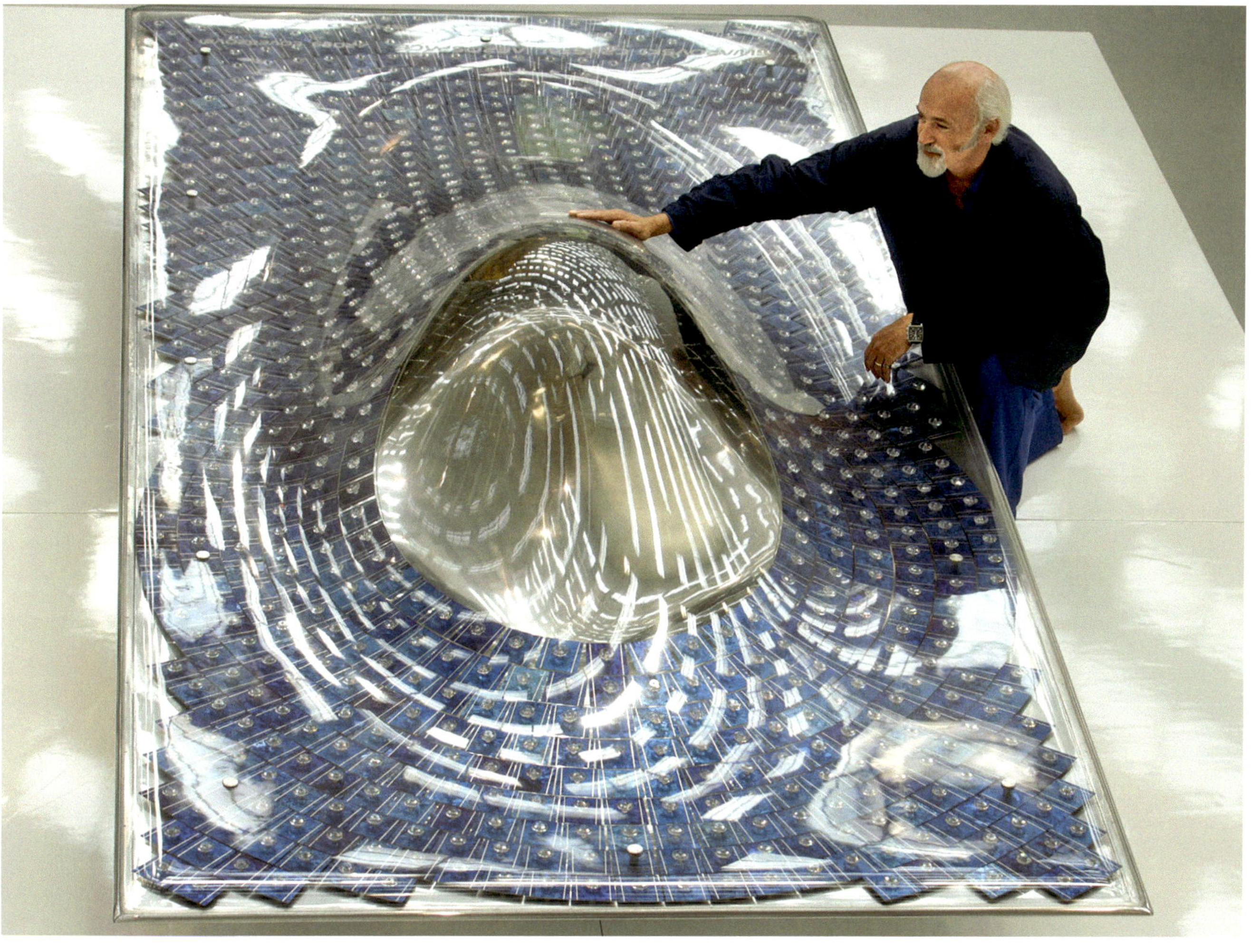

technology. Today, as images of space are becoming ubiquitous in the media thanks for instance to SpaceX, bringing this level of beauty and logic to everyday, terrestrial life would surely be a most relevant and wonderful thing.

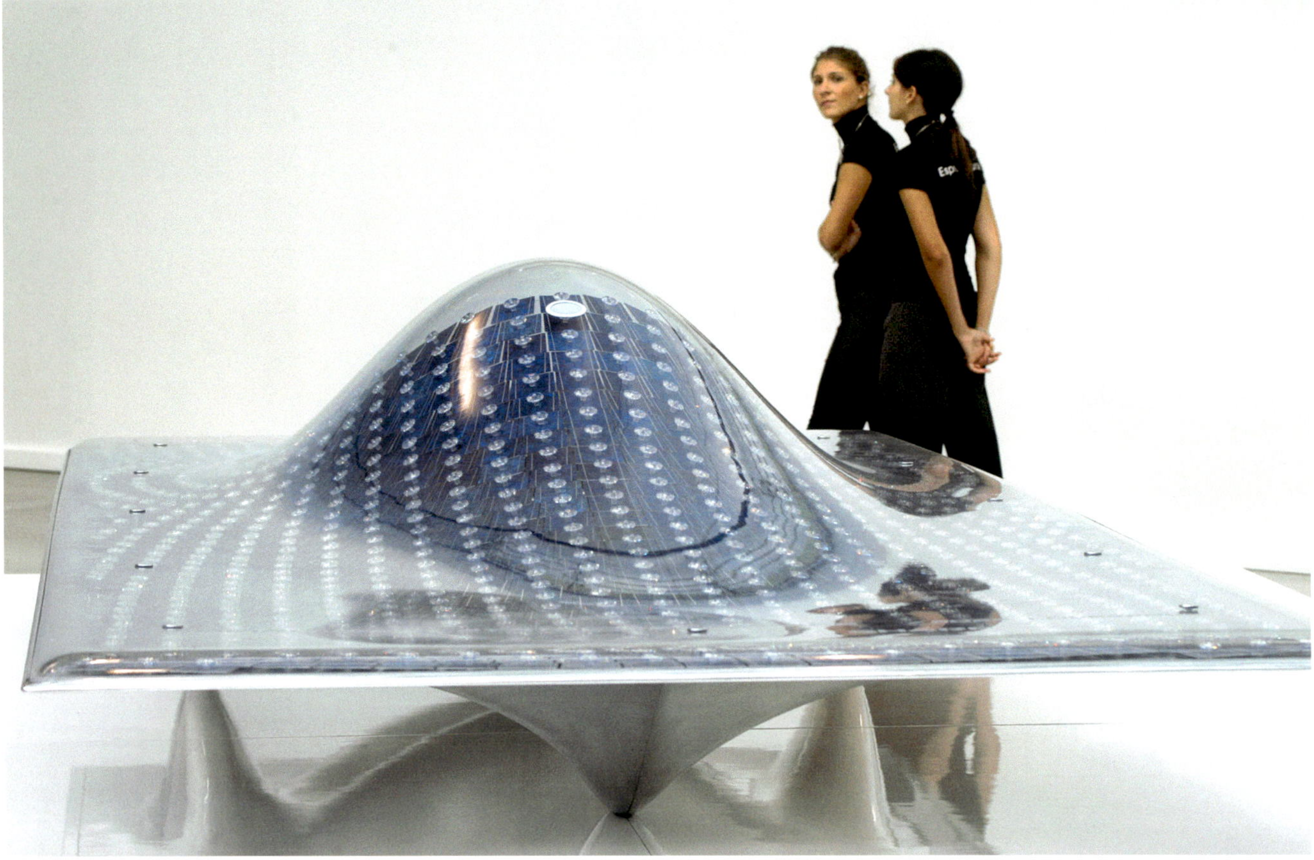

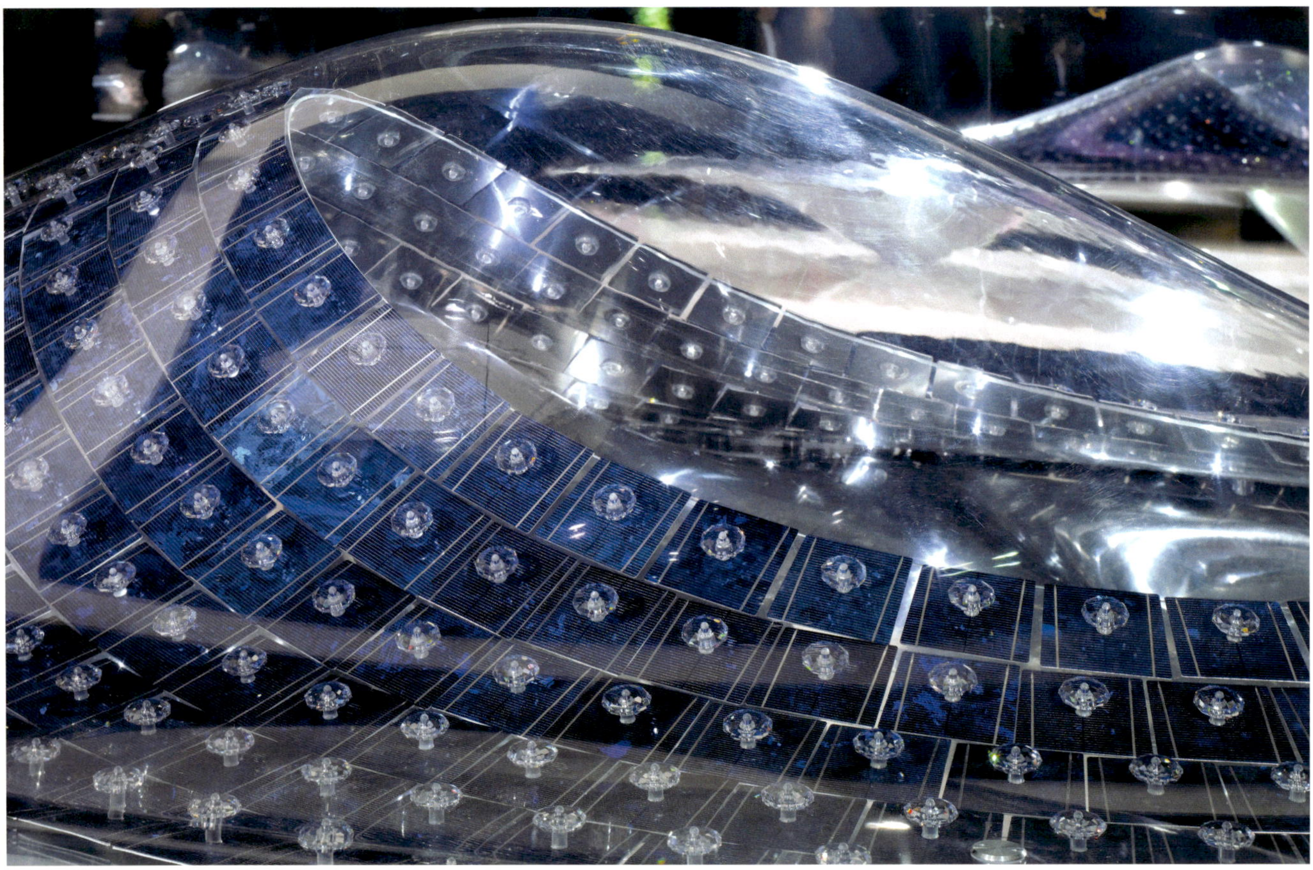

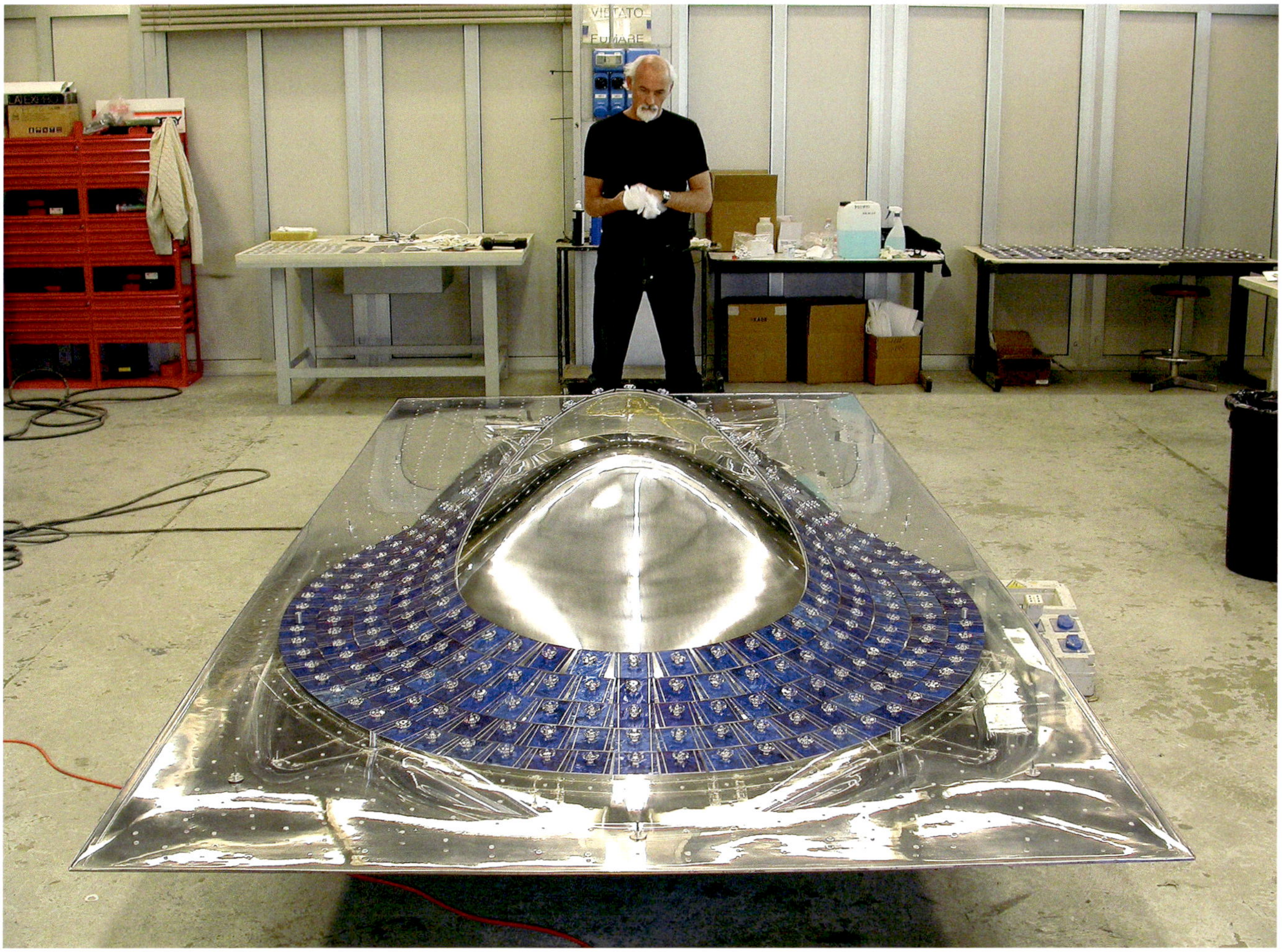

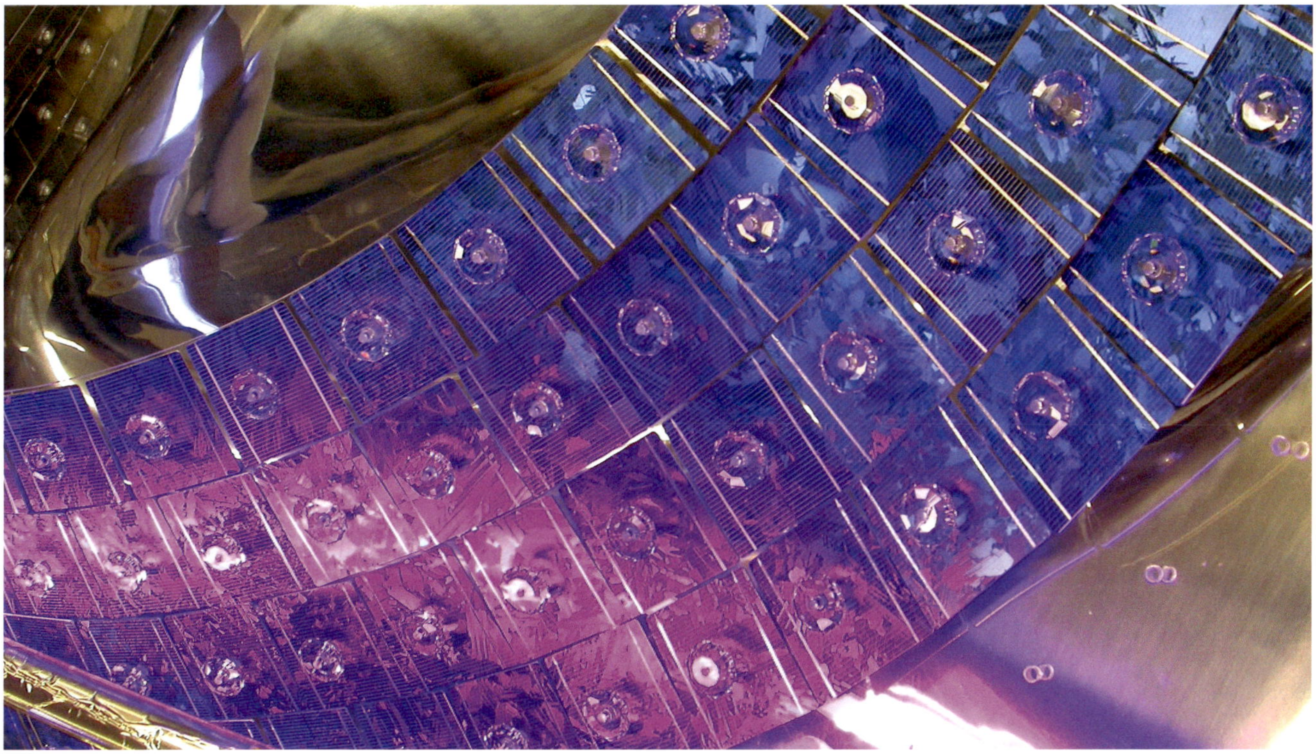

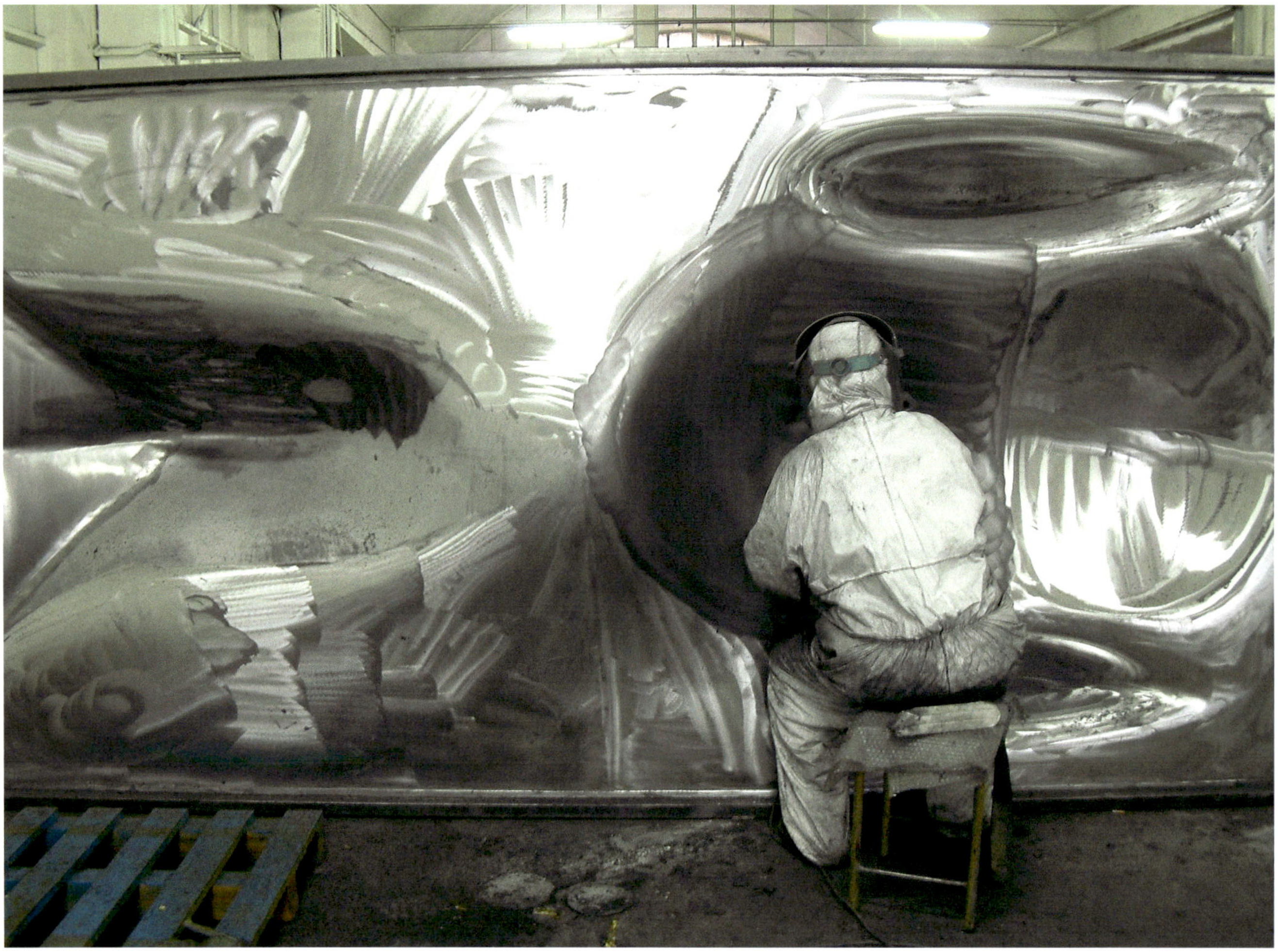

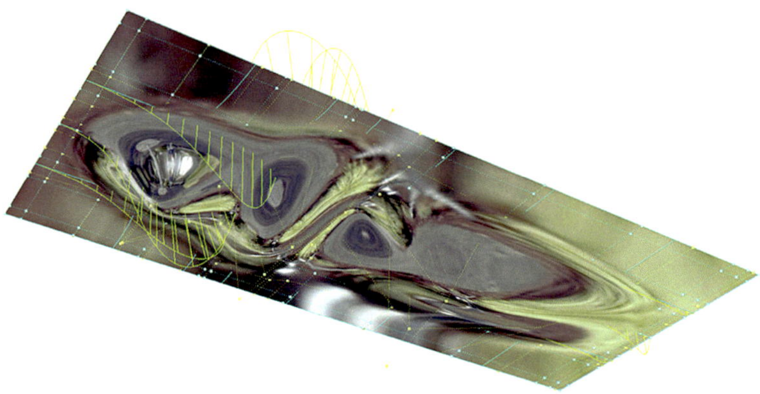

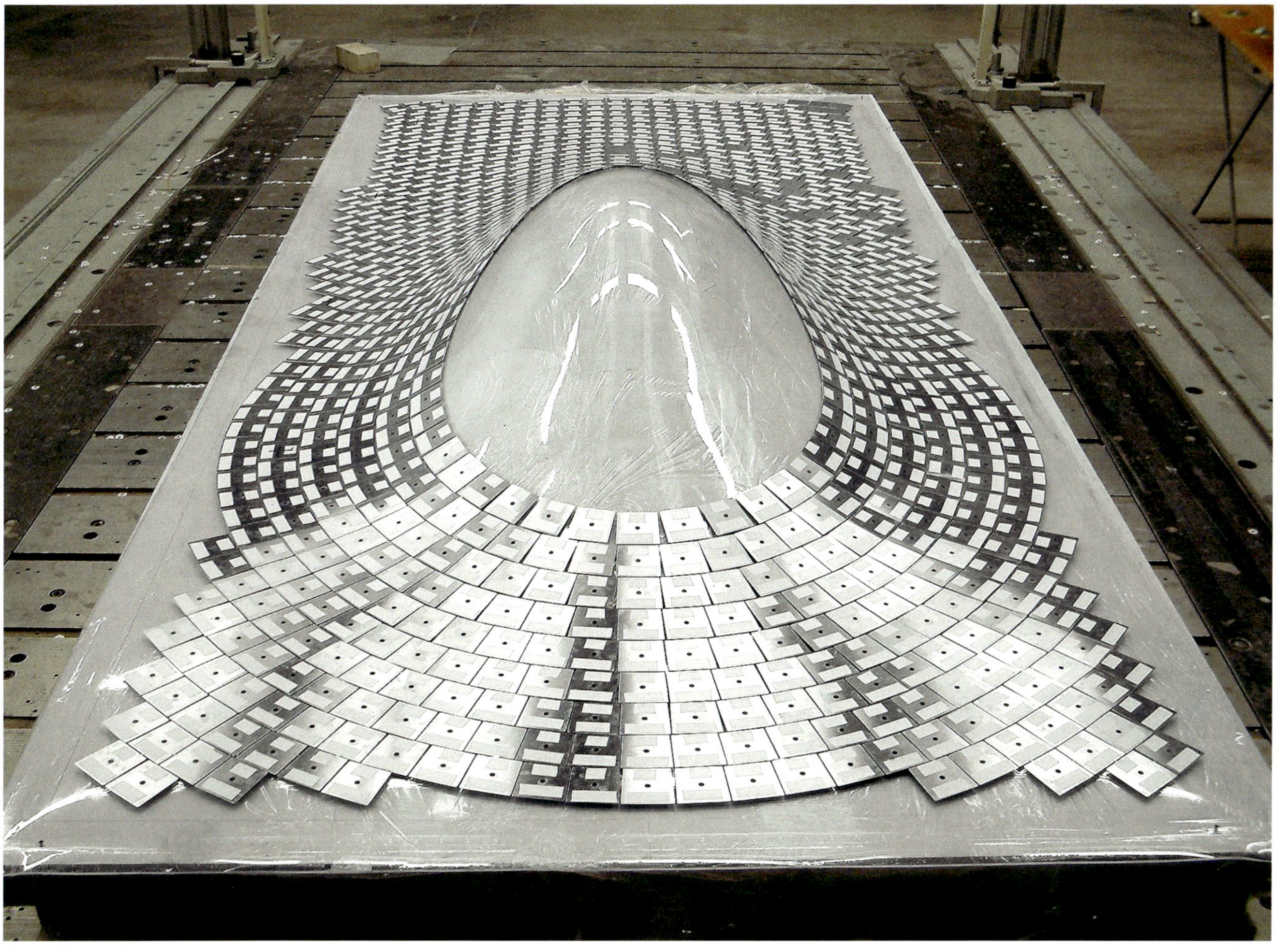

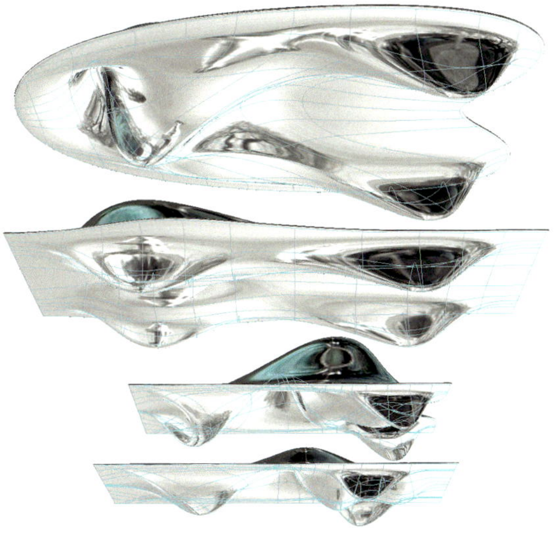

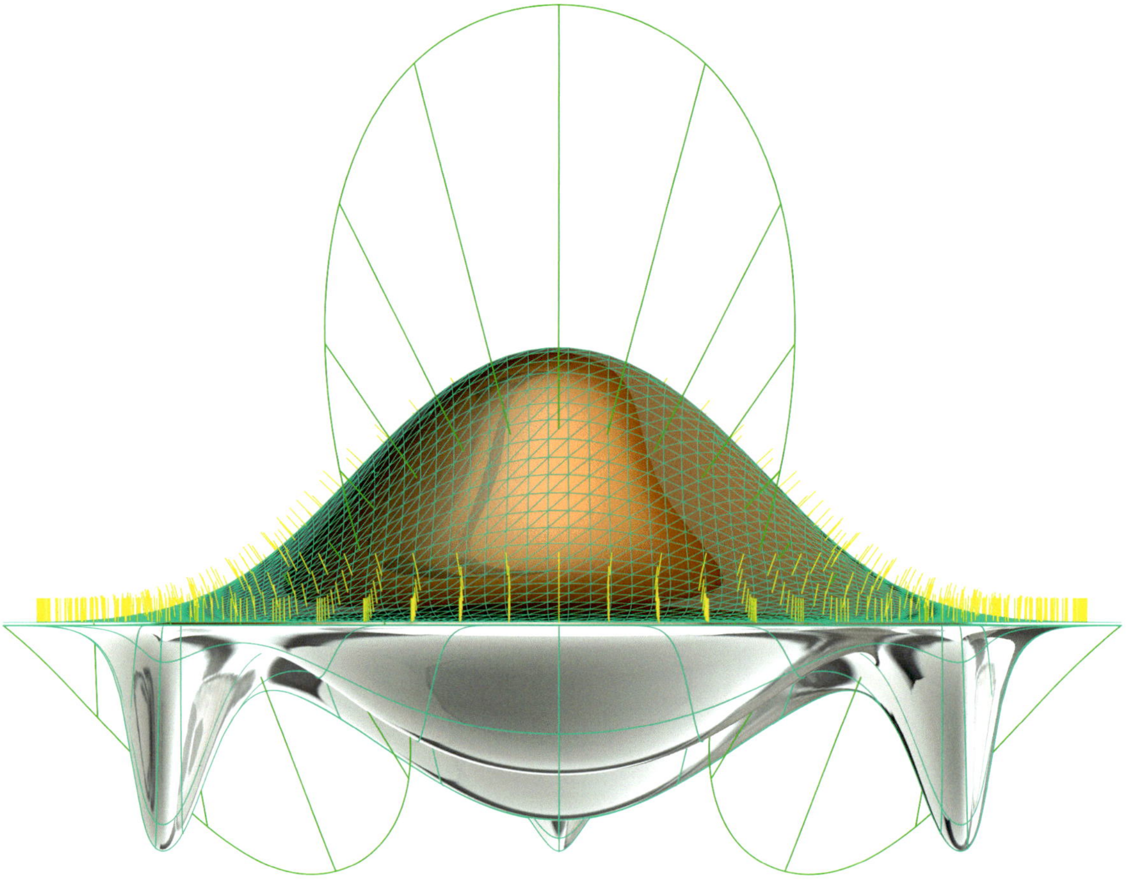

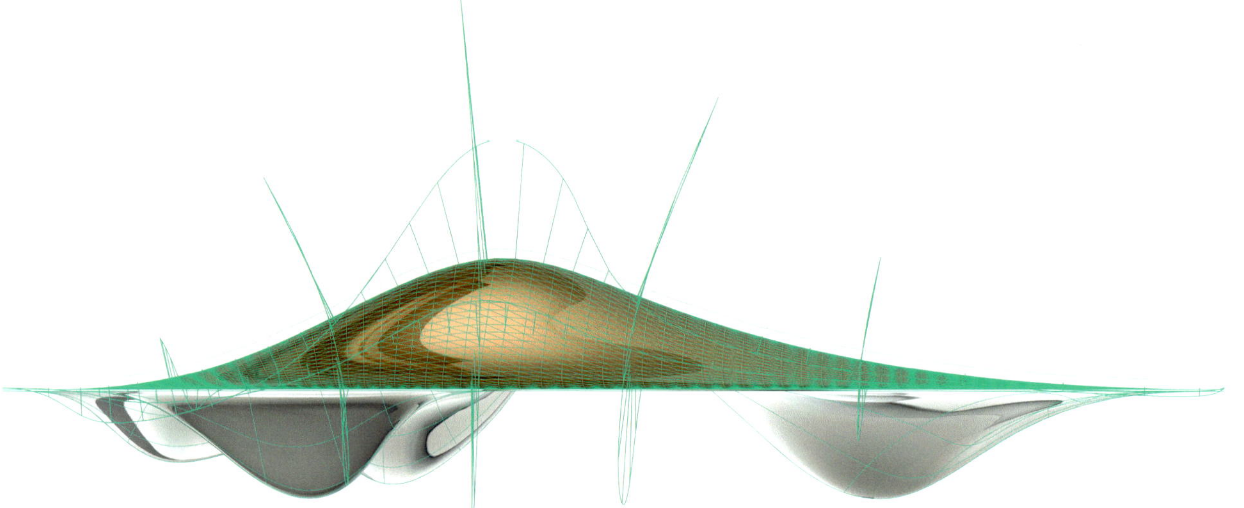

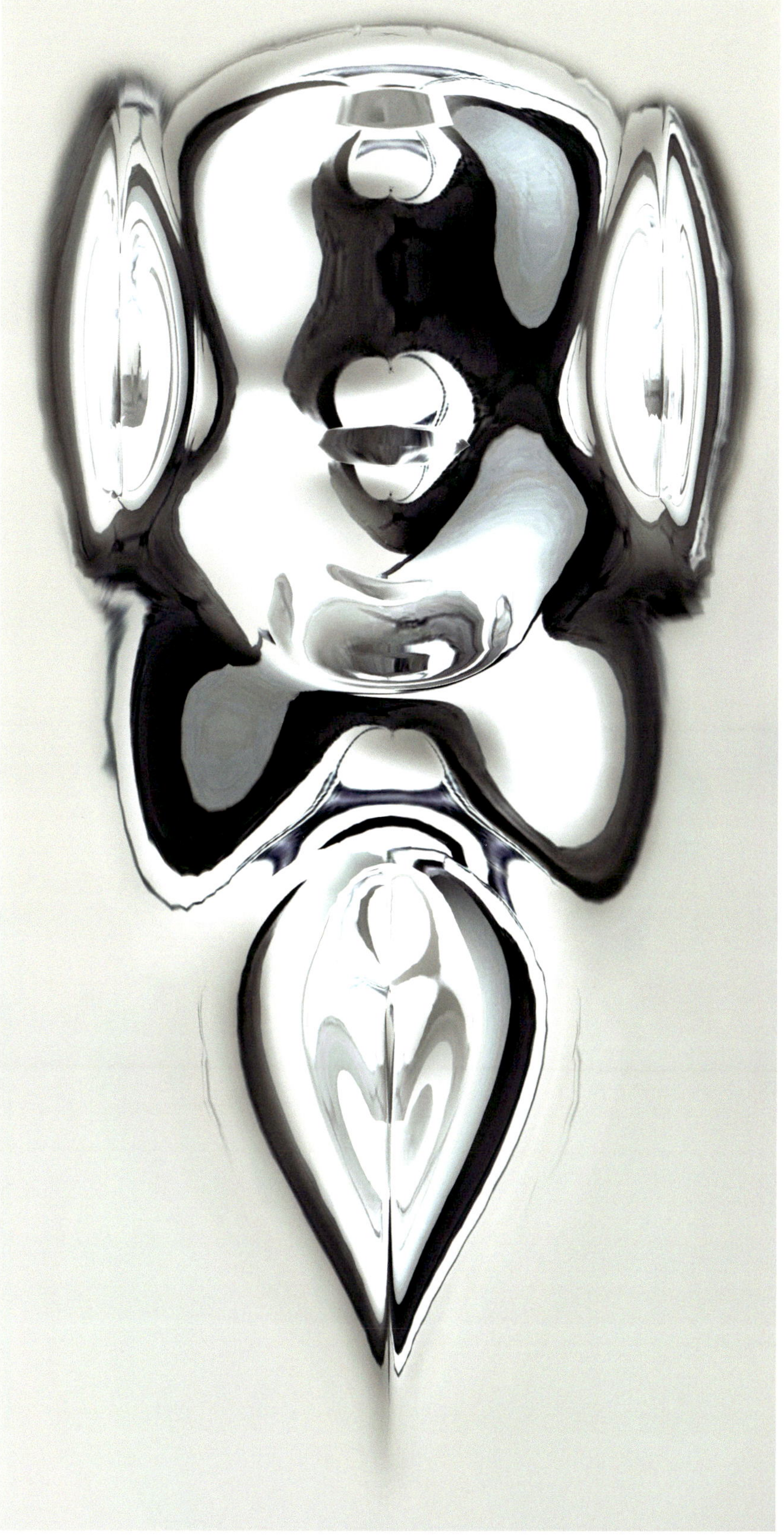

Twin'Z electric car

2012–2013
Prototype
Mixed materials; 152 × 363 × 170 cm
Manufacturer: Renault (France)

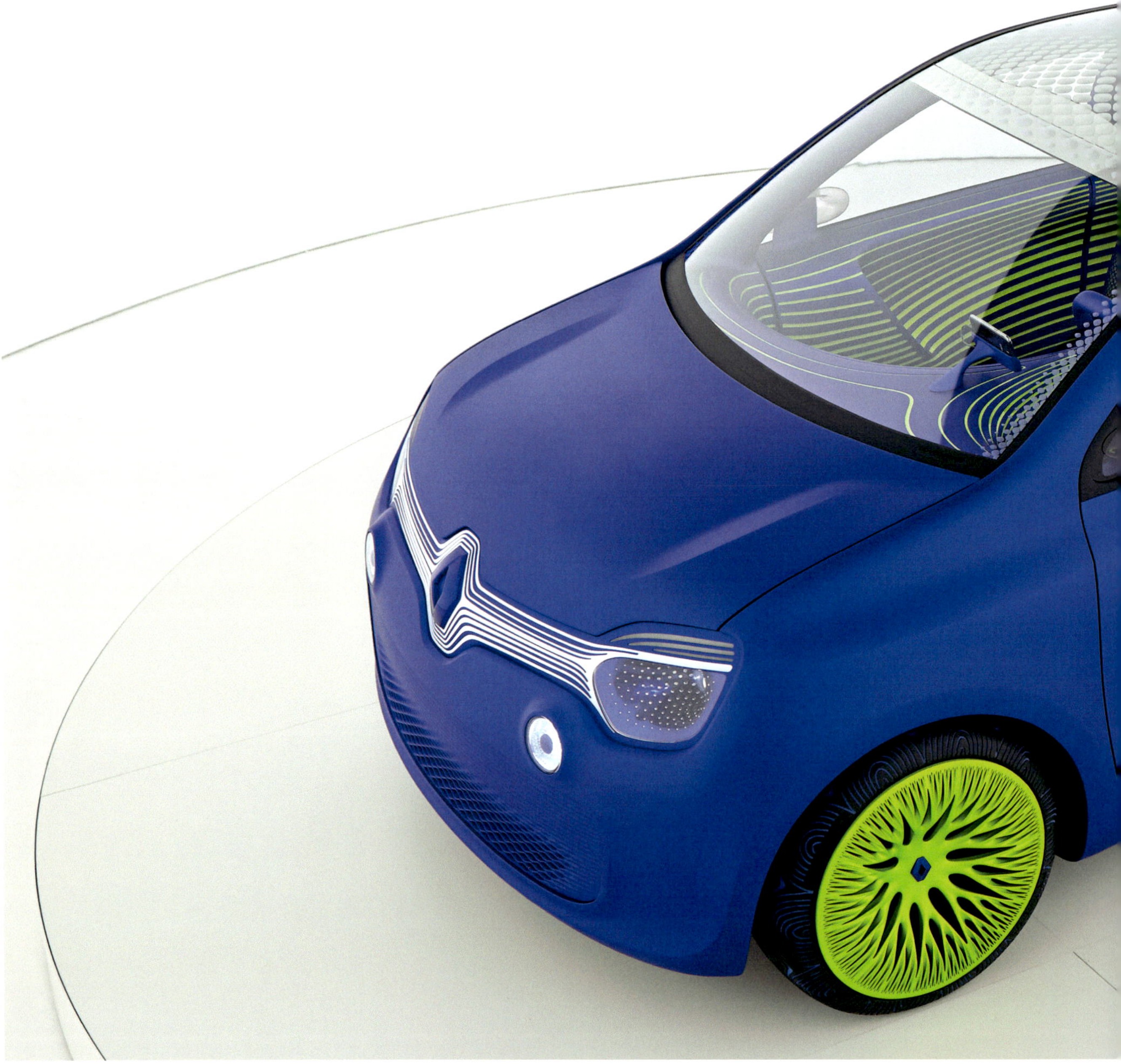

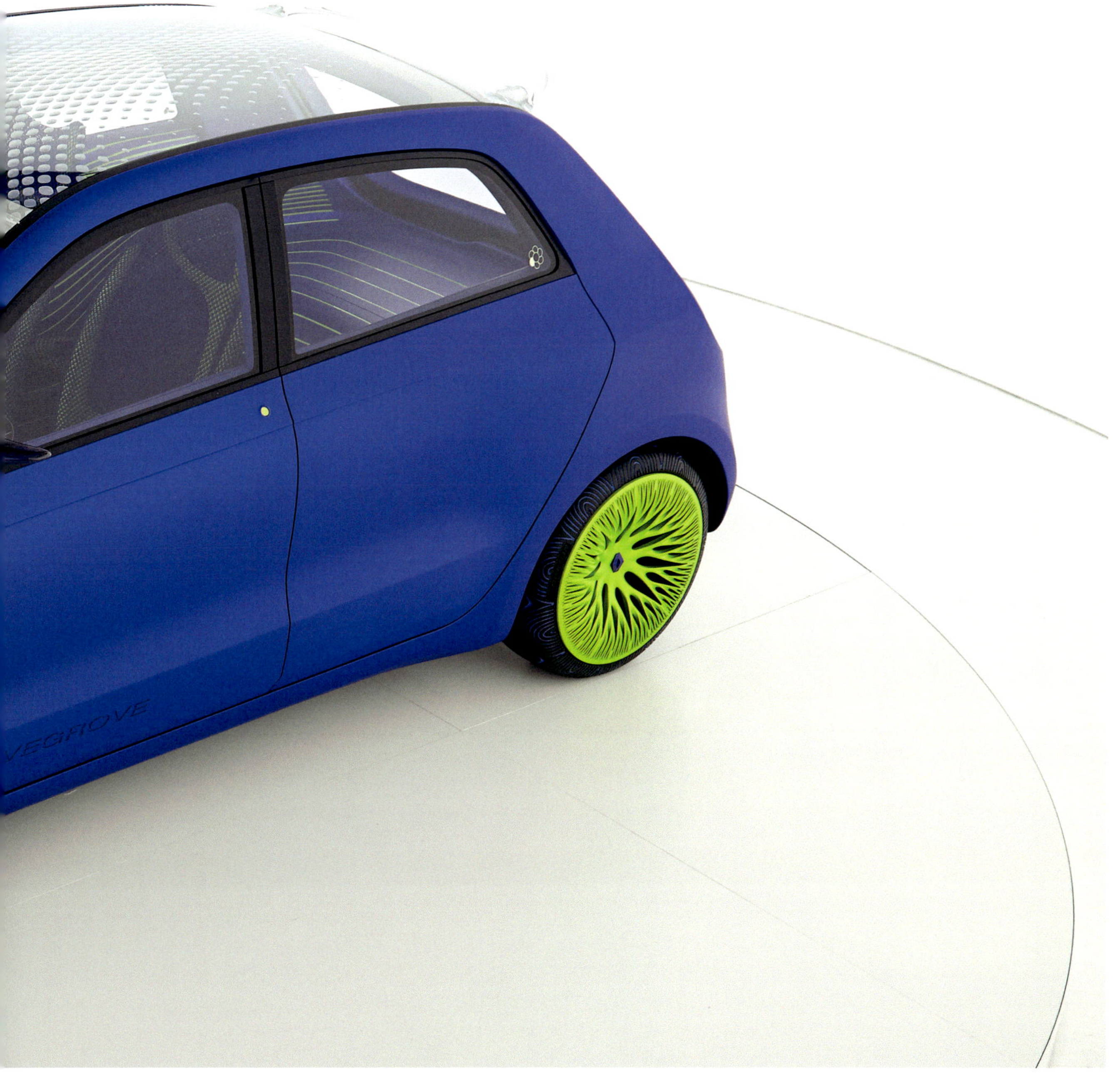

Twin'Z electric car

Ross Lovegrove transformed the funky curves of the 1990s Twingo by thinking up the Twin'Z, an electric, rear-wheel drive, city car. Shown for the first time in 2013 at the Triennale Design Museum in Milan, this fifth concept car to emerge from the Renault design strategy stunningly demonstrates an encounter between the worlds of furniture and cars. Starting from the basic outline devised by Renault Design, Lovegrove was given complete freedom to outfit the bodywork and to design the passenger interior as a dematerialized technological cocoon, combining a natural and fluid formal vocabulary with the latest 3-D modelization techniques. Its modern and playful appearance stimulates emotions and plays with the passenger's senses in relation to the notion of travel. A result obtained by an unprecedented use of light, color, and texture, it makes the Twin'Z an expressive, almost living object.

"Lightening the car by physically dematerializing it," that was Ross Lovegrove's intention. Being 3.62 meters long and weighing 980 kilograms, the vehicle attains a top speed of 130 km/h, with a range of 160 km. This car is equipped with rear-hinged doors, and both the front and rear doors open electrically with a simple gentle touch. Furthermore, the elimination of the central B-pillars allows the whole interior space to be opened up. In a pared-back style, the passenger compartment is covered with a network of fluorescent green bands, symbolizing the energy flow and emphasizing the contours. The framework of the carbon fiber seats, which seem to grow naturally from the floor, lend an ethereal quality to the entire passenger compartment, equipped with innovative digital features. The seats are covered with a blue fabric woven from a light, 3-D mesh that is comfortable as well as waterproof and fireproof. The dashboard disappears, replaced by a simple touch-screen tablet controlling the car's settings and a smartphone indicating the speed.

"Passengers are hooded in a technological envelope that bathes them in a light that responds to the energy and pulse of Twin'Z," explains Lovegrove. "The interior is not broken up into separate components, and all the passengers feel very much part of the traveling experience," he adds. Furthermore, the glass roof increases the feeling of space. A network of electroluminescent diodes (LED) runs through it, which diffuses the light that falls harmoniously up

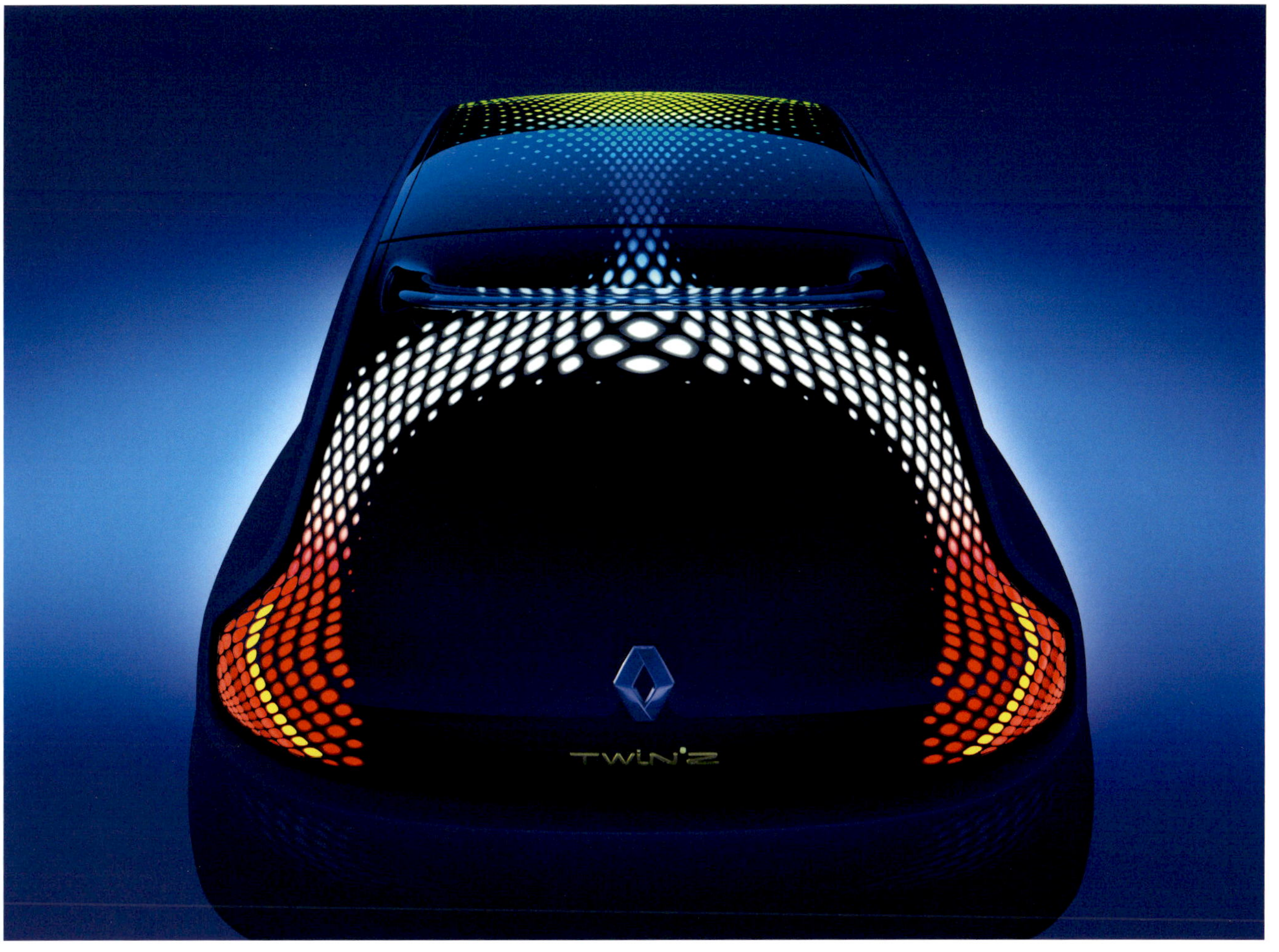

to the rear window. Designed to create a vortex effect, the front grill directs the air flow and maximizes the aerodynamic performance by softening the wind resistence. It takes its inspiration from the iconic models of the brand, such as the Renault 5 and the Twingo. The smooth, matt finish of the bodywork is applied like a skin, made iridescent by the natural aspect of the coloring and the reflected light that highlights its "electro-natural" appearance. The car's color is inspired by International Klein Blue (IKB), the hallmark of the artist Yves Klein. "It echoes France's cultural heritage while also mirroring the virtues of our planet," the designer remarks. Even the headlights demonstrate the organic nature of the vehicle. The diurnal driving lights have been given fibrous radial spokes, which look like the iris of an eye, while the main headlights, accentuated by "eyelids," give the Twin'Z an anthropomorphic appearance. Finally, the wheels were designed as a complete entity by combining parametric modelization with 3-D printing.

Twin'Z electric car

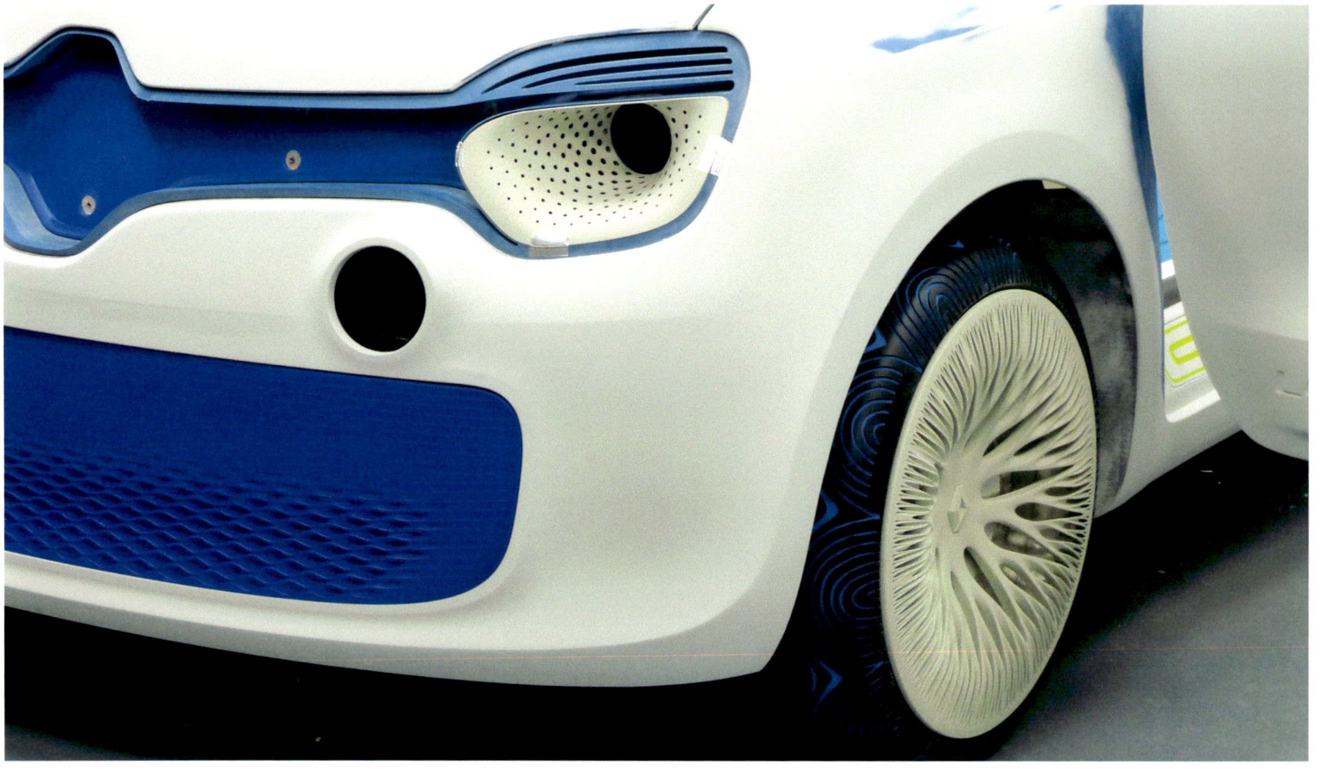

The wheel rims fan out from a central point, spreading like branching vegetation in a luminescent green color, meeting at its edge the tire devised by the Michelin company. According to Lovegrove: "The car has become the symbol of progress and civilization, a symbol of man's ability to master cutting-edge technology and to transform materials into high-precision objects, accompanying our daily lives. The use of composite and recycled materials today opens up new opportunities." The Twin'Z frees itself from the traditional codes of representation thanks to digital modelization and encapsulates the necessary transition "from the 'orthodox' mechanical aesthetic . . . to the ecological principles of a 'softer' aesthetic" in light of the dangers threatening the environment and the current overconsumption of energy.

MS

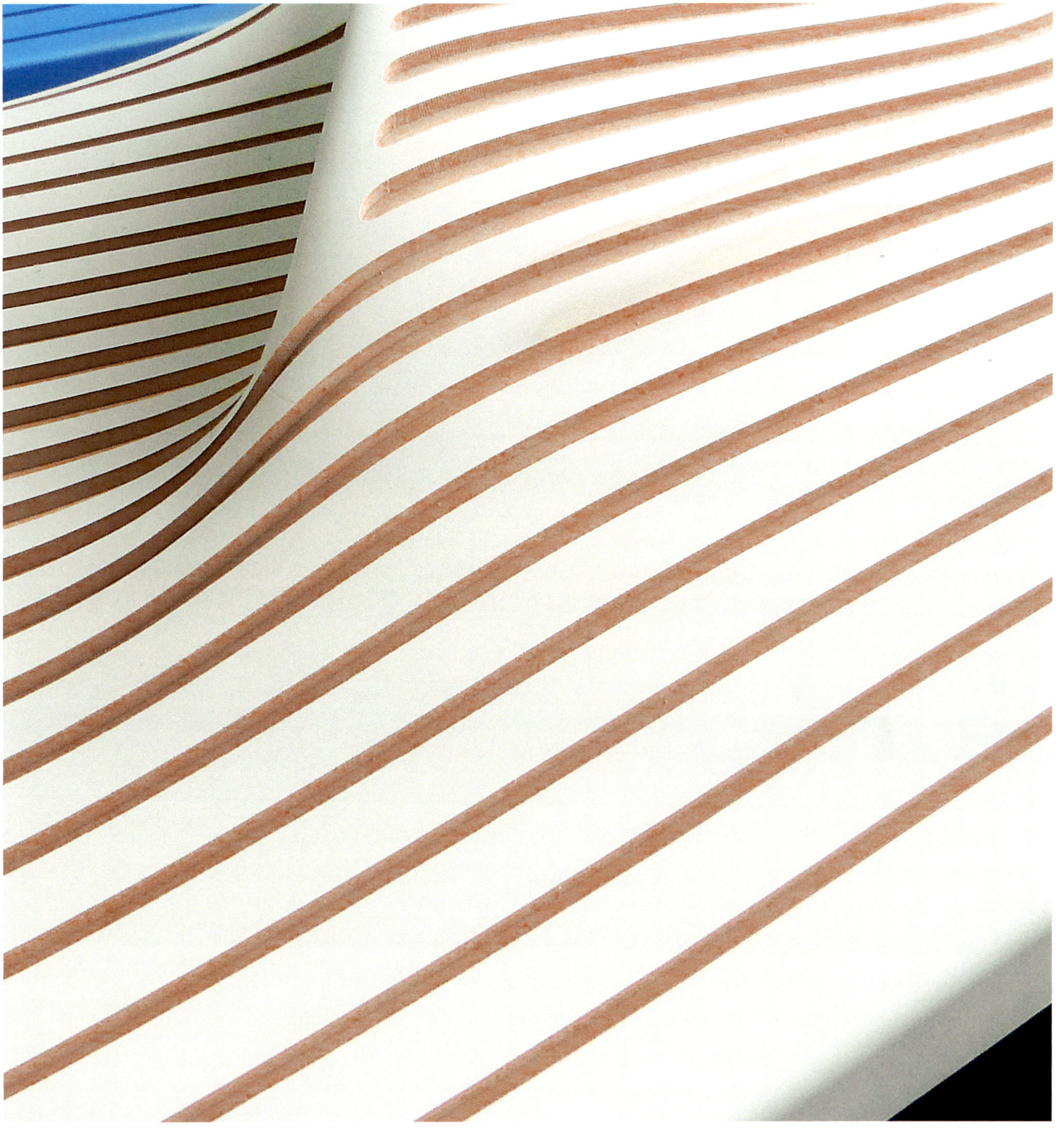

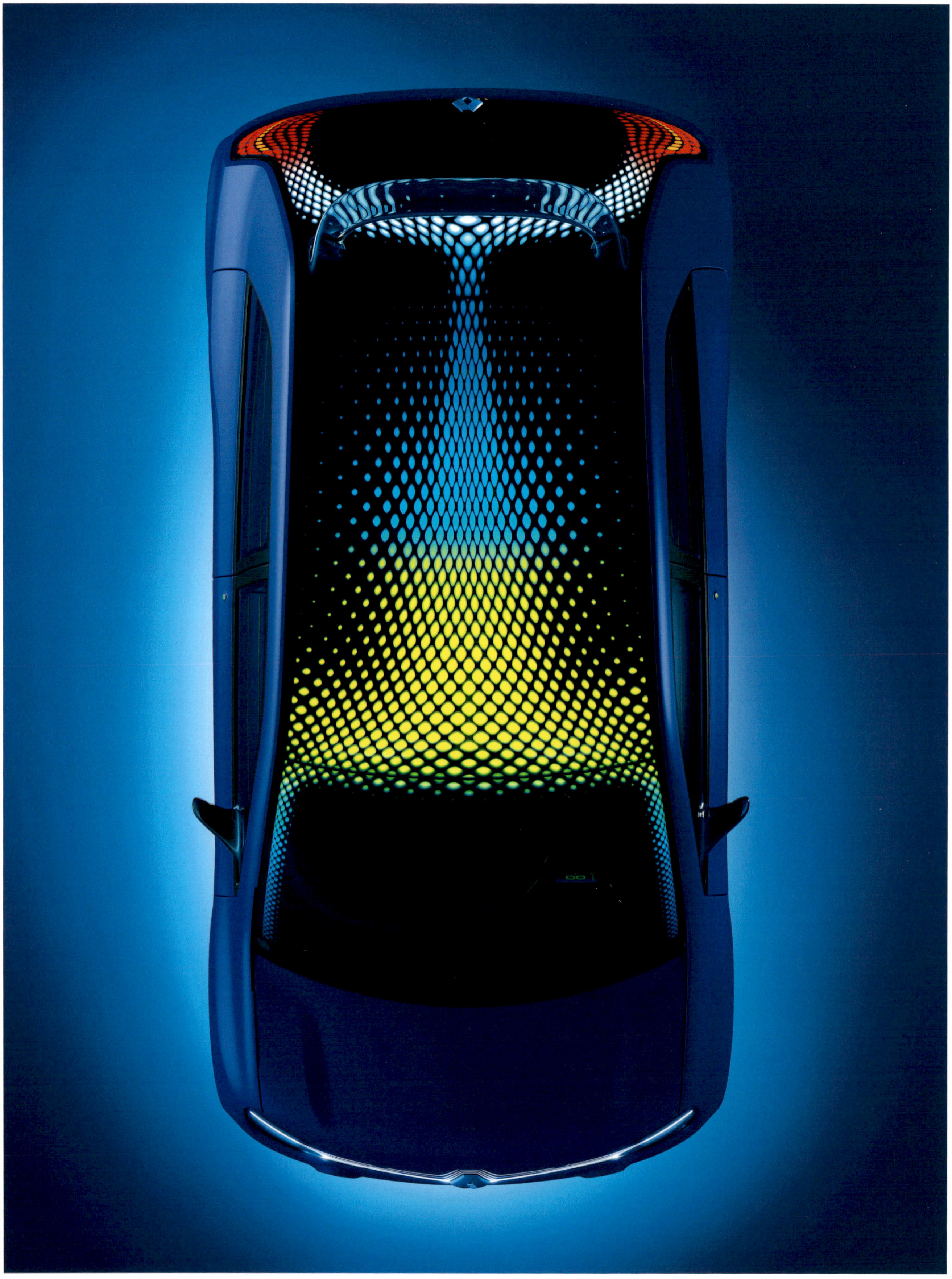

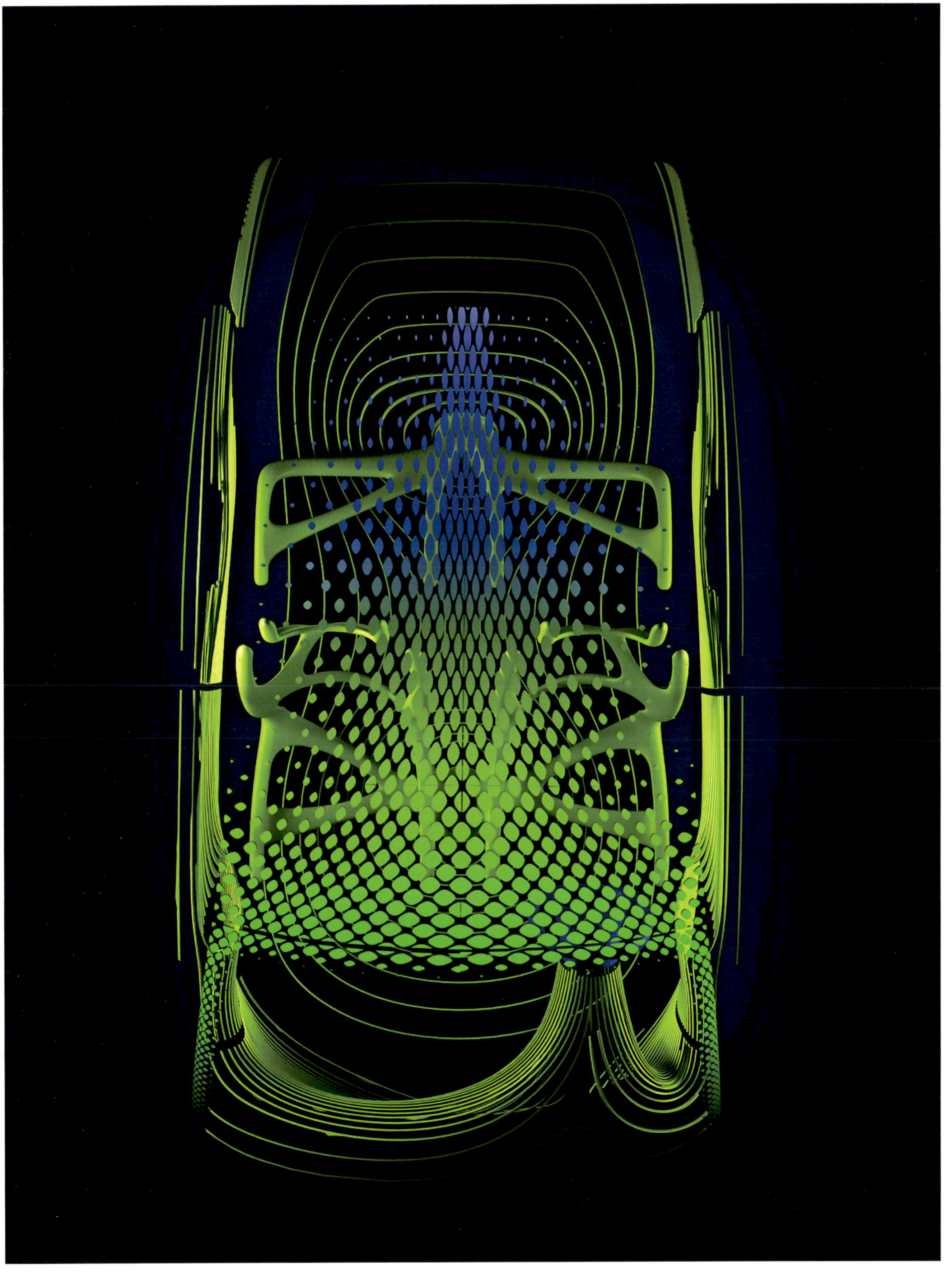

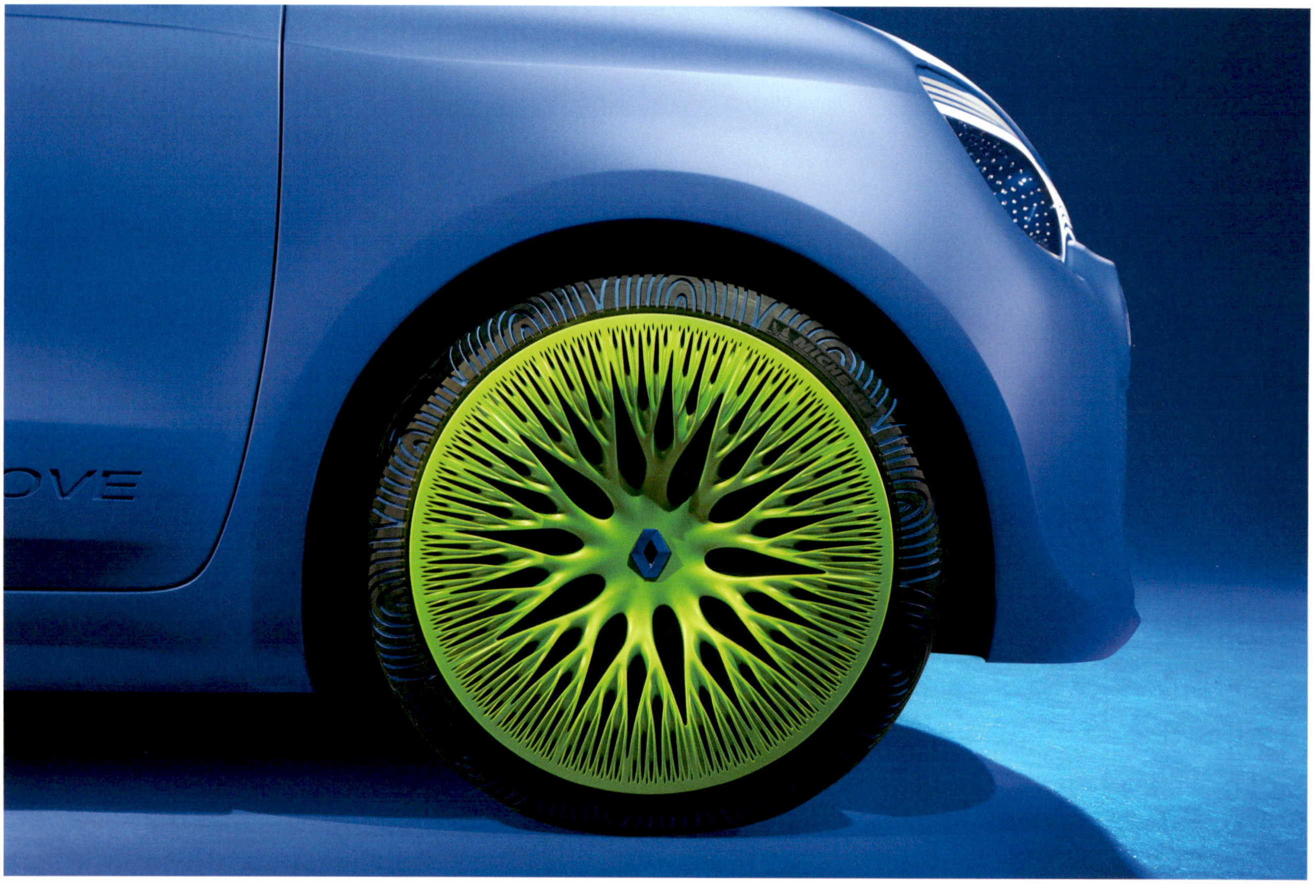

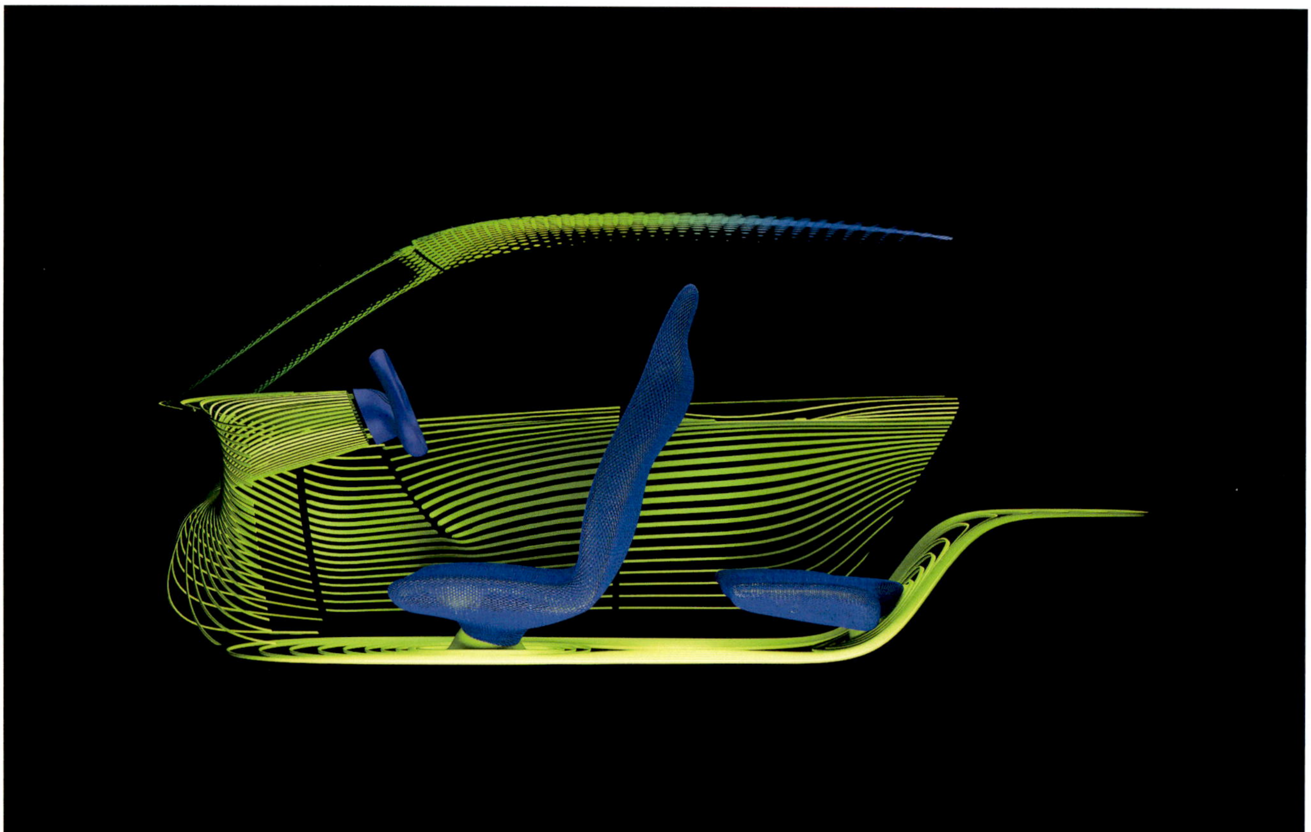

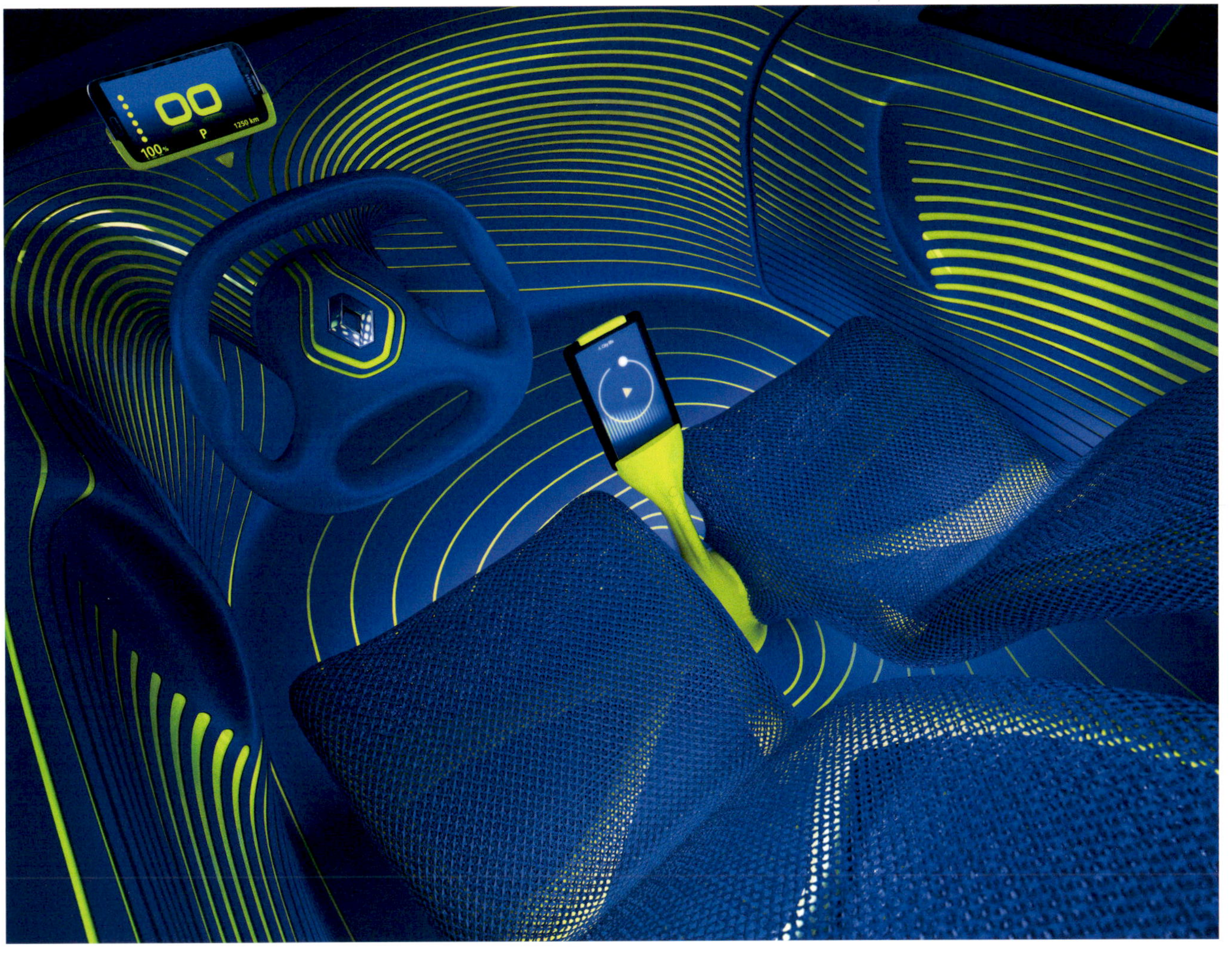
100%
P
1250 km

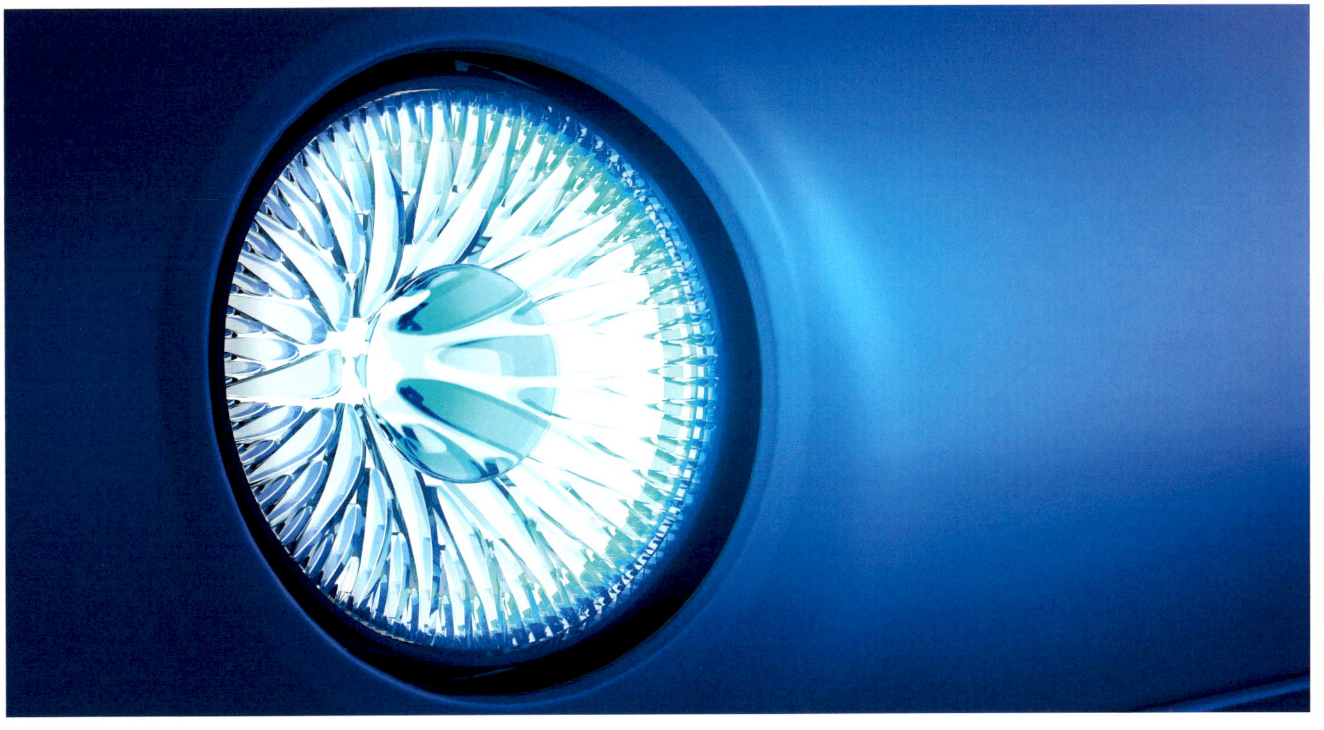

Lasvit LiquidKristal pavilion

2012
LiquidKristal glass, steel framework,
Barrisol stretched ceiling; 400 × 1200 × 800 cm
Manufacturer: Lasvit (Czech Republic)

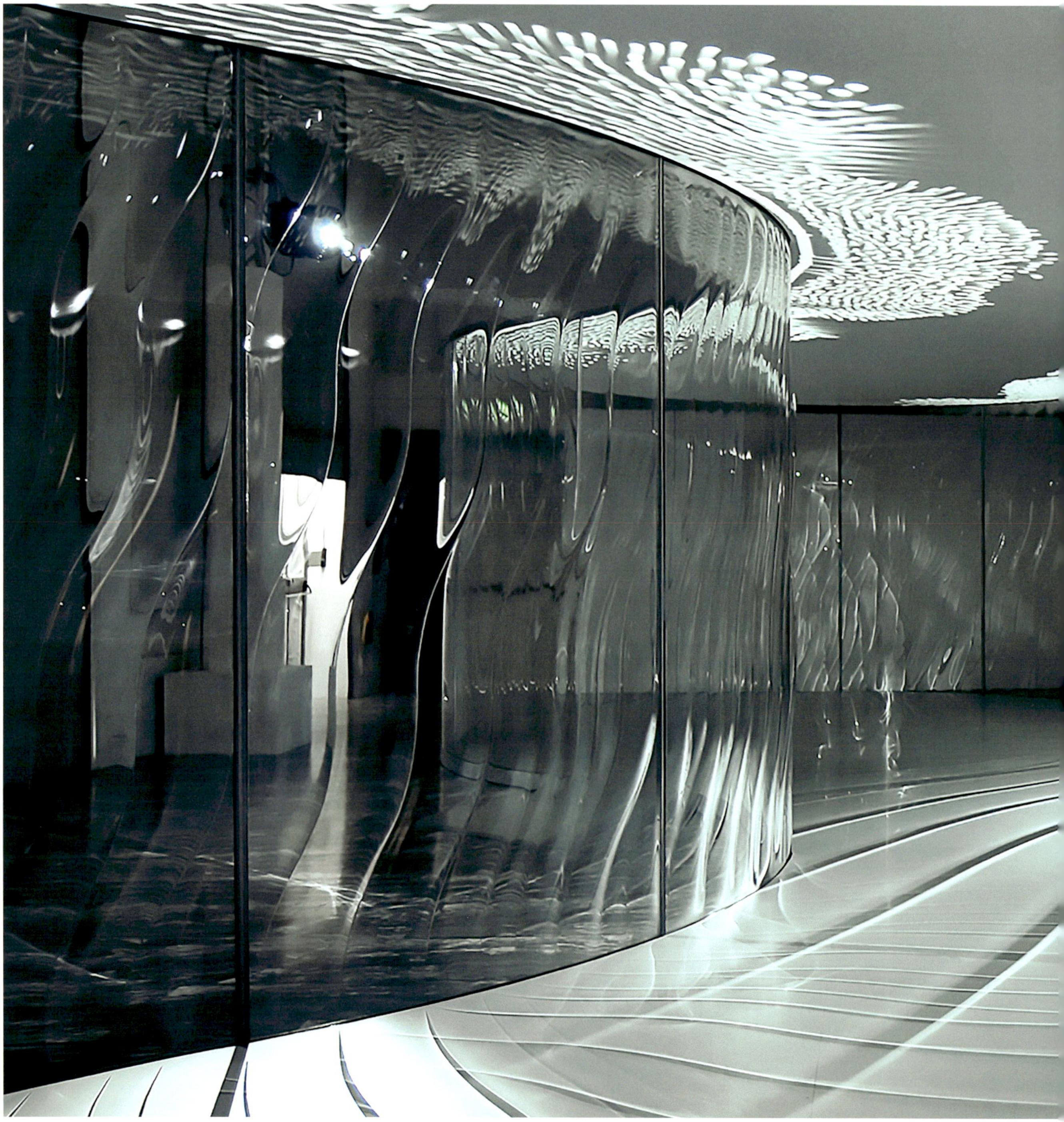

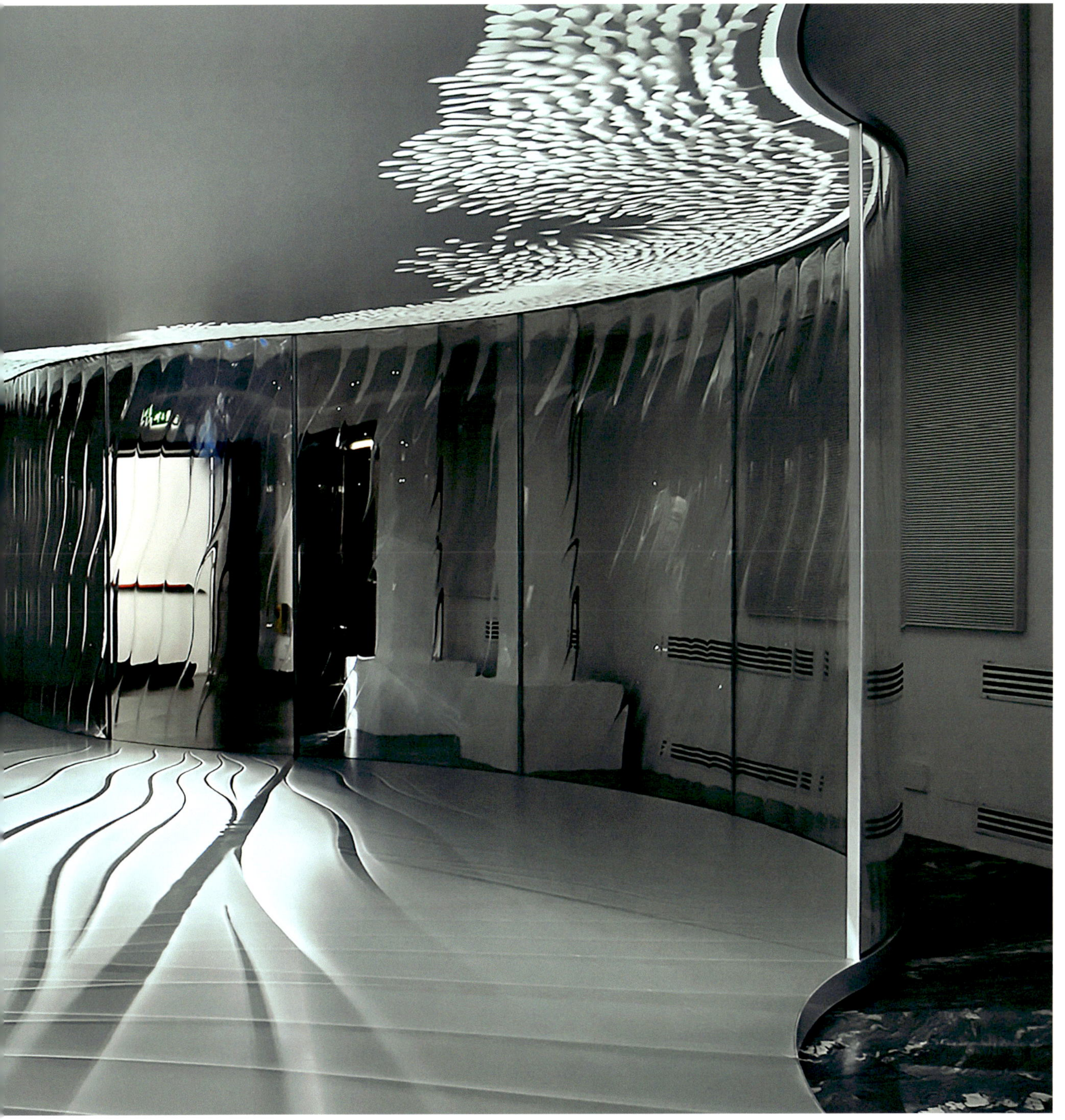

Lasvit LiquidKristal pavilion

In 2012, at the Triennale Design Museum in Milan, Ross Lovegrove, in collaboration with the Lasvit company, presented the LiquidKristal pavilion, an undulating glass structure that invites visitors to immerse themselves in space. Lovegrove worked for more than a year with the Czech glass manufacturer to create wavy surfaces that appear to pass straight from a liquid to a solid state. Approached following the principle of convergence between design, physics, and technology dear to the designer, the pavilion is arranged into three sections of twenty-four free-standing glass panels over a total surface area of eighty square meters. On the structure's ceiling the visitor can discover projections by Biothing (Alisa Andrasek), which show digitally generated moving forms inspired by nature. At the same time the undulations in the internal surface of the panels create optical distortions, changing the visitor's perceptions as they look toward the outside.

Lovegrove digitally explores the large-scale itemization and consolidation of models found in nature. Here, following the concepts of "organic essentialism" and "fluid dynamics," simulation software and digital tools have allowed the designer, using mathematical models, to simulate the behavior of glass subjected to controlled thermic induction. For LiquidKristal, Lasvit developed an innovative procedure that sped up the natural process of cooling glass. The accelerated transition from fusion temperature to cooling, obtained by an extremely precise regulation of the temperature, allowed the beautiful optical effect of reflections, as seen through water, to be arrested in the material. The firm then created a system of flexible molds to capture these glimmering effects, which repeat and flow in a continuous manner, from one panel to another.

Poised at the boundary between dematerialized architecture and a digital environment, the pavilion introduces a completely new approach to glass, in which its qualities of transparency, lightness, and the refraction of light are emphasized.

MS

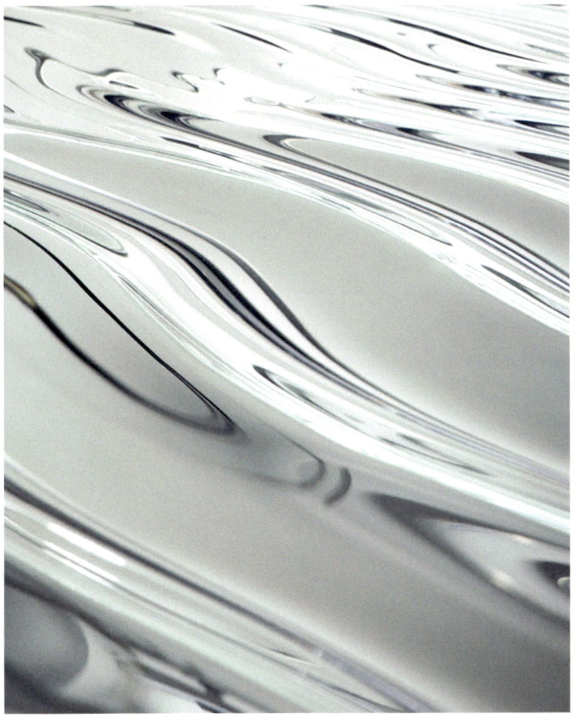

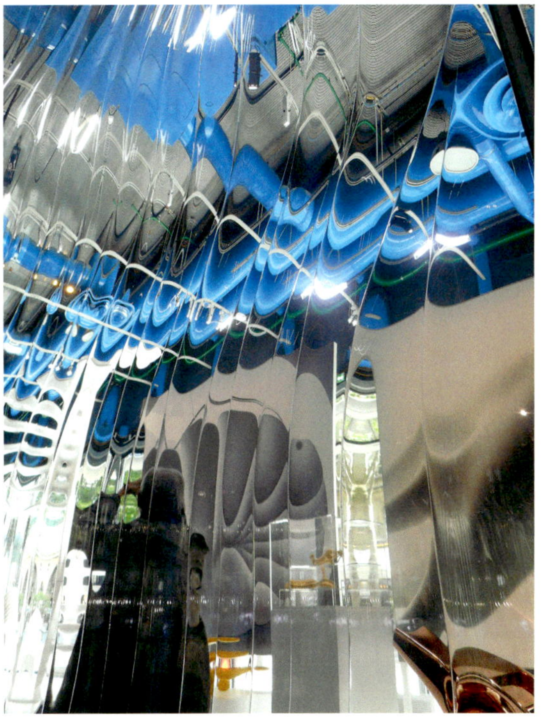

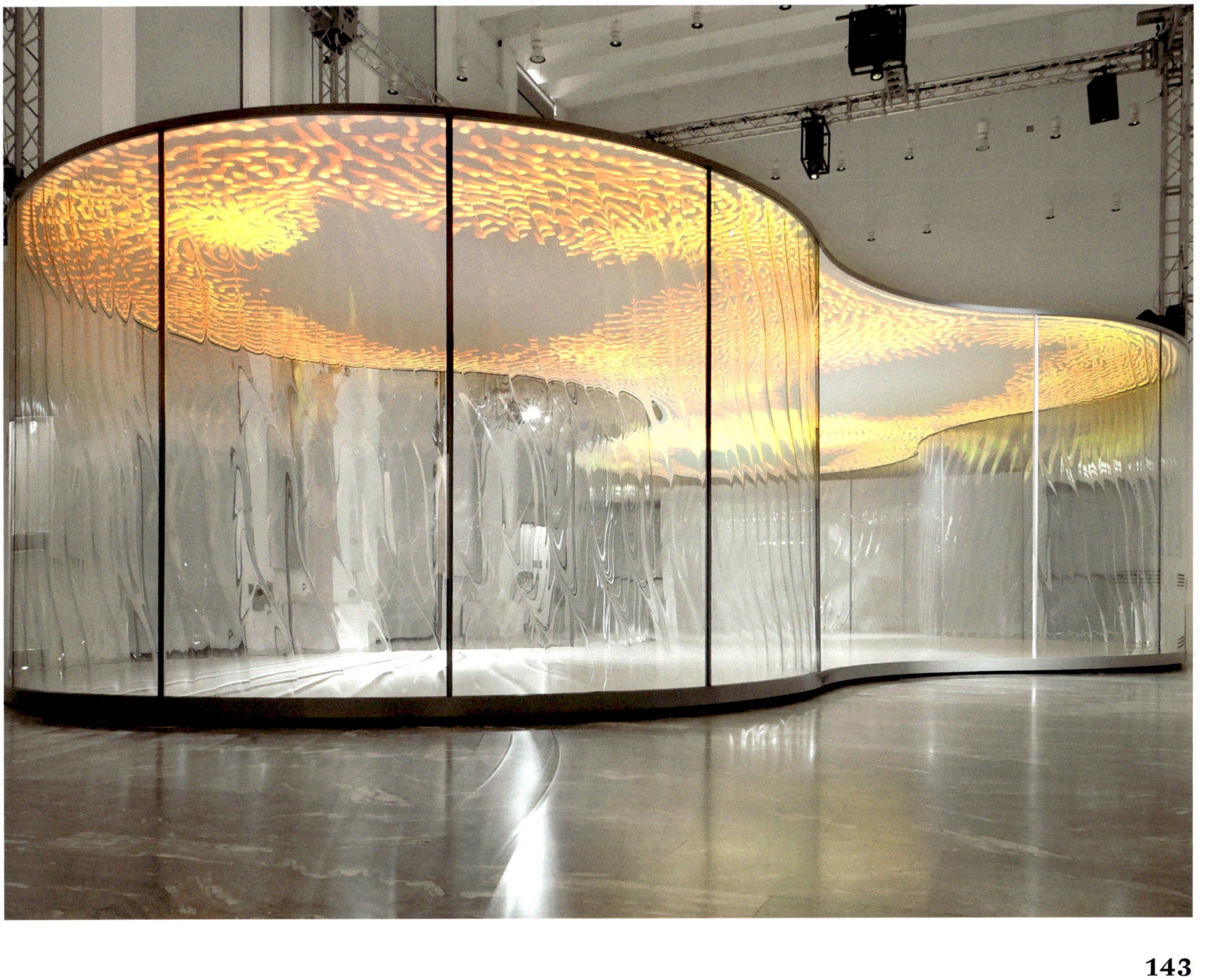

Barrisol Cocoon

2016–2017
Sheets of Barrisol Mylar on an aluminum structure;
265 × 840 × 840 cm
Computer-generated images of the project
Manufacturer: Barrisol Normalu (France)

Specially conceived for the present exhibition, *Barrisol Cocoon* is an installation that serves as a display arrangement. The cross-shaped structure was designed with Barrisol, the worldwide leader in the technique of stretched ceilings. It relies on the principle of a double skin: an exterior skin in stretched Barrisol canvas, with metallic gold reflections, and a black interior membrane, in a light-absorbing fabric, to cover the central space pierced by trapezoidal openings. Plunged into darkness inside this delicate pocket, *Instinctive Override* (see p. 147) is a hybrid installation, ranging between a futuristic concept car, reduced to its elementary form of a water droplet, and a videosculpture with hints of optokinetics. It was created in partnership with the Italian engineering laboratory G-Studio and shown for the first time in 2012, during *Futuristic Primitivism,* the title of the twenty-third Biennale Interieur, in Kortrijk, Belgium. For the Centre Pompidou, Ross Lovegrove has created an installation that functions like a 3-D screen, from which digital animations designed by the architect Alisa Andrasek are projected onto an anamorphic fiberglass object. Inspired by the shapes of large marine mammals, perfected in the process of a long evolutionary history, the ovoid surface is washed over by an uninterrupted dynamic flow of luminescent filaments, like the hypnotic visuals of aerodynamic properties. This effect is underscored by the soundtrack, which, played simultaneously, corresponds to the rhythm of the visuals.
Inspired by the concept of "organic essentialism," Lovegrove defines the "prelinguistic" fluid forms adapted for tomorrow's mobility, which will take place at supersonic speeds. An ideal aerodynamic vehicle or electro-organic form in the making, *Instinctive Override* symbolizes the moment at which human instinct overrides scientific analysis. "What this achieves," concludes Lovegrove, "is a sense of motion, lightness, and almost aquatic serenity, a gentleness not felt when standing next to a contemporary car, where intuition and emotion are suppressed by the physical complexity, weight, and often cold aggression of its construction."

MS

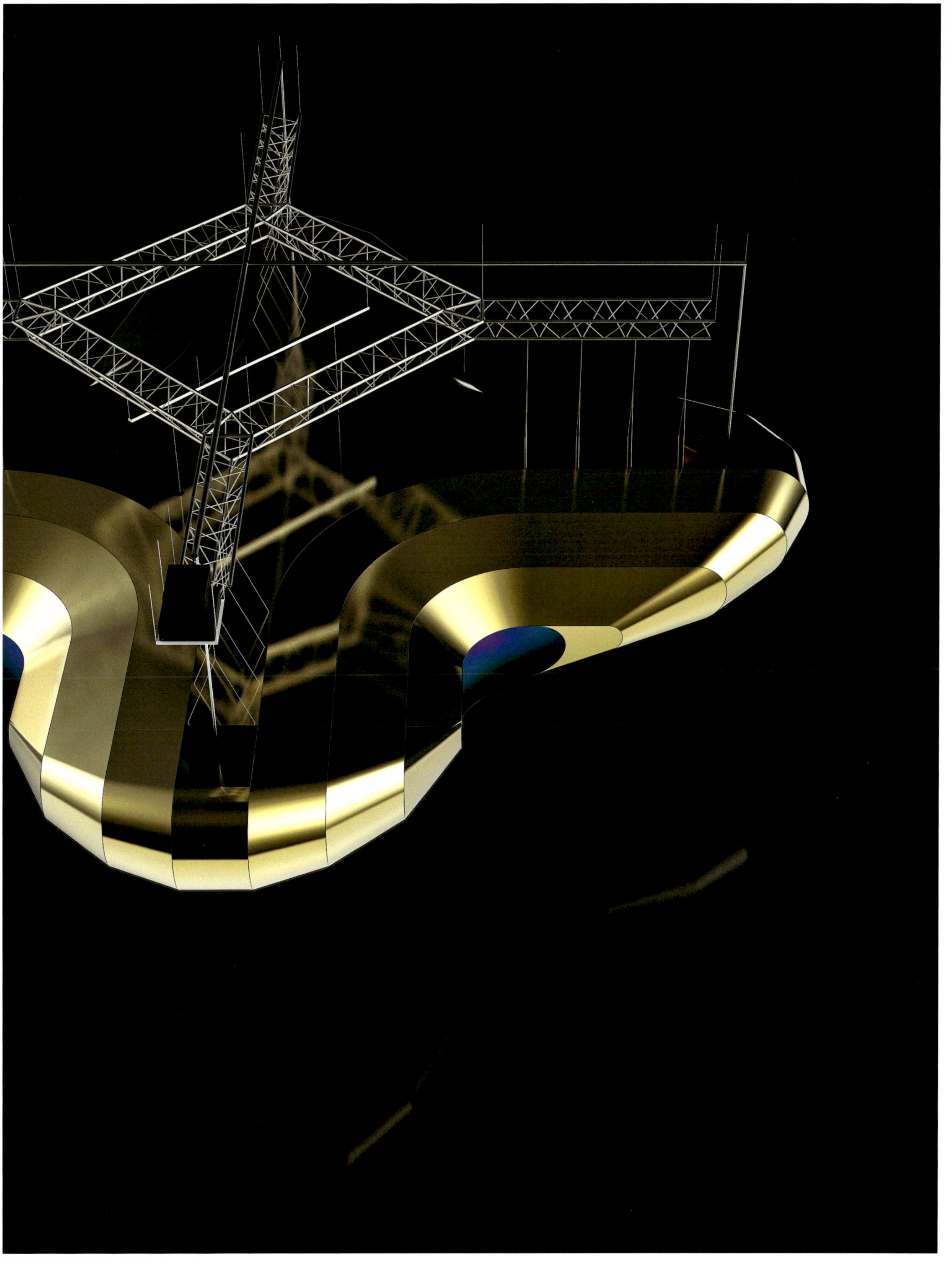

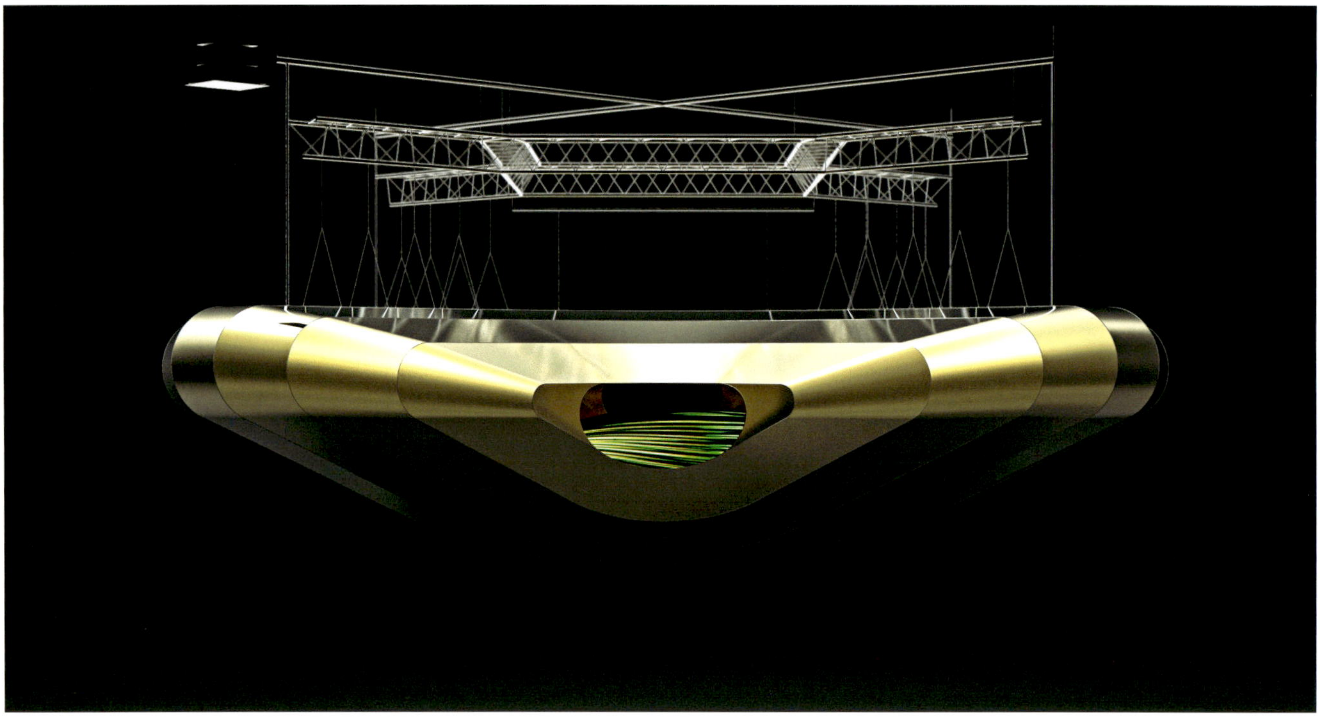

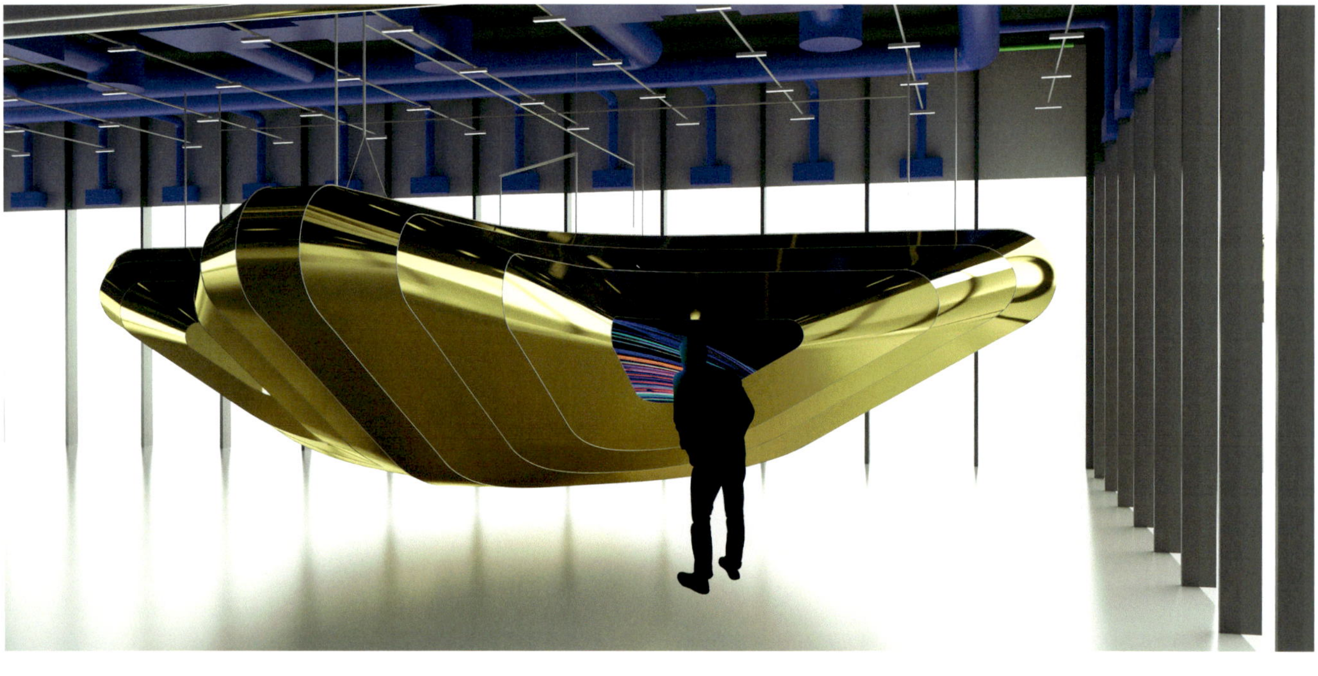

Instinctive Override

2012
Installation: video projection on fiberglass form;
126 × 406 × 167 cm

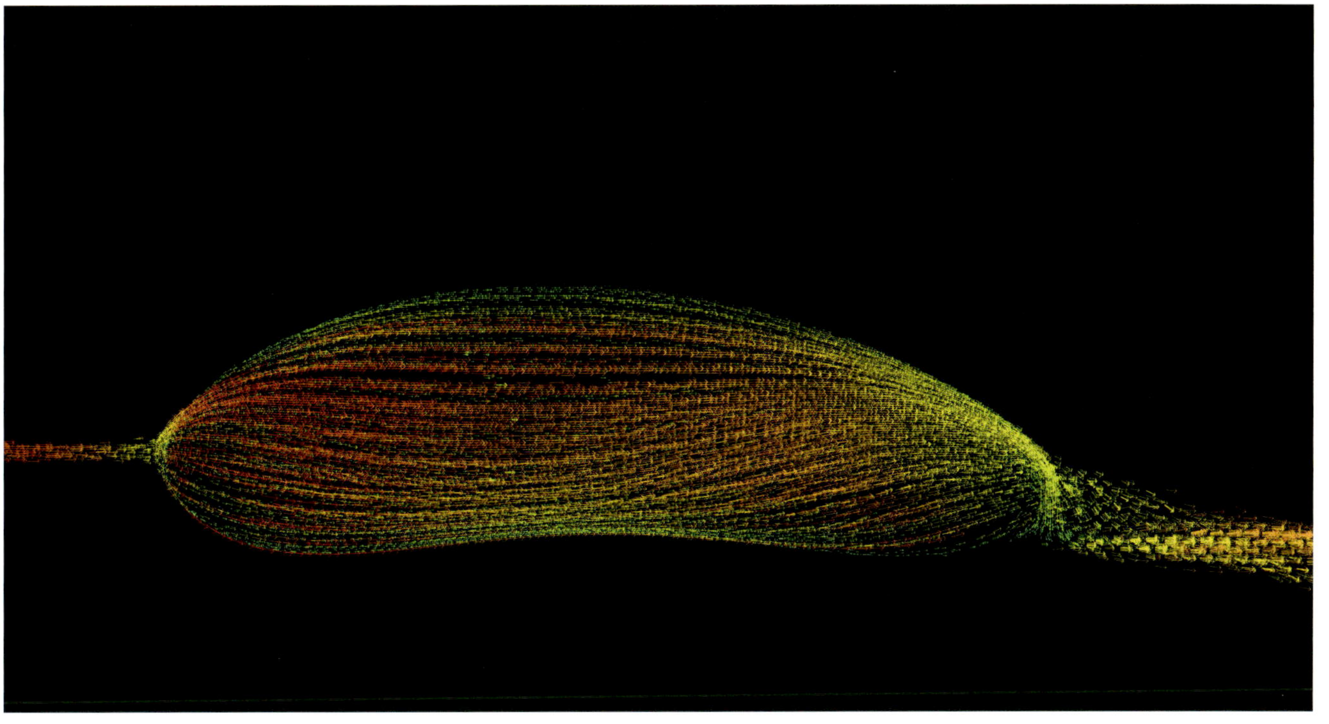

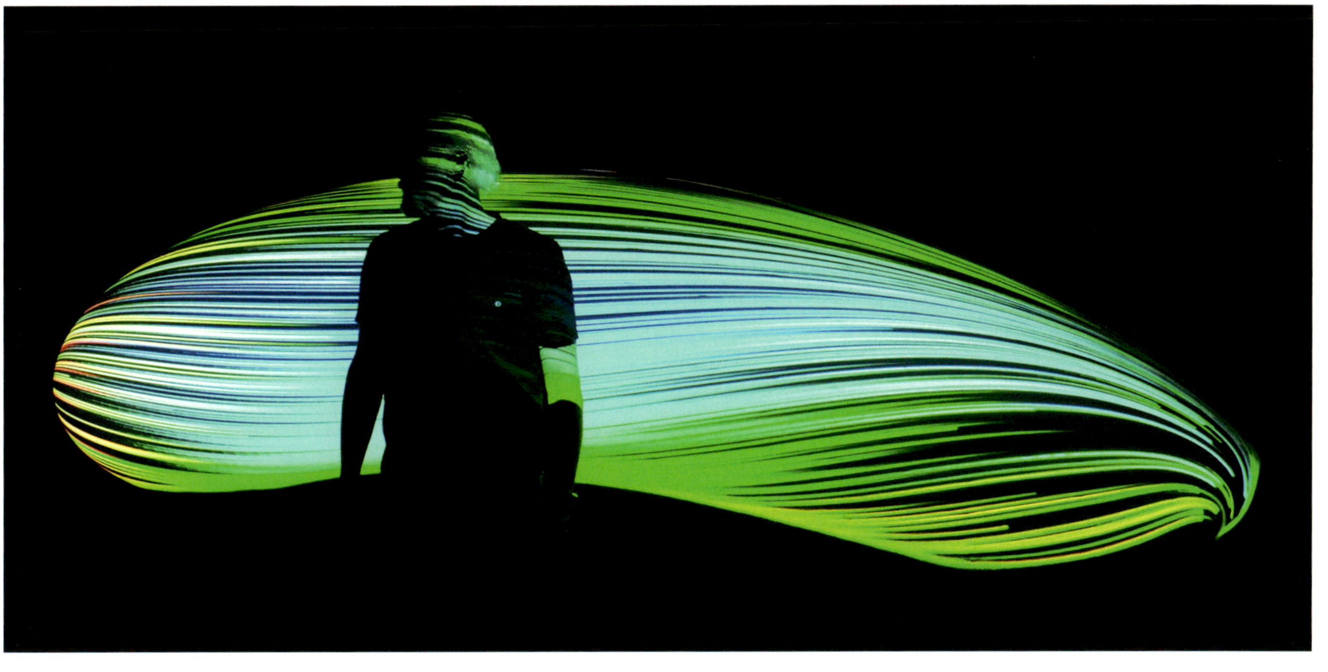

From Analogue to Digital

Ross Lovegrove

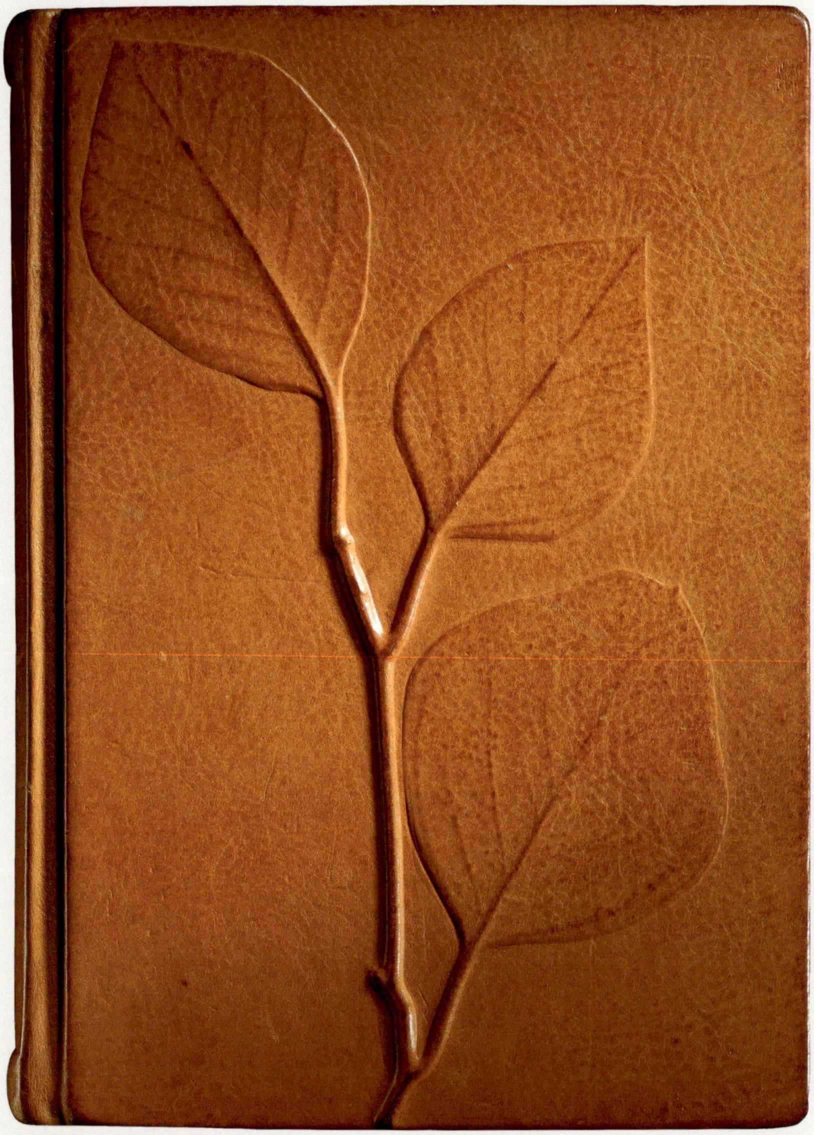

Above

Notebook with leather binding over a dried plant, 44 × 16 cm

Page 150

Research drawing with a text written "in the manner of Darwin" during his voyages at sea. (Running short of paper, he had to use the same sheet at different angles.), 2016, pen on paper, 42 × 29.7 cm

Studies for the roof of the Renault Twin'Z, 2012, pen on paper, 42 × 29.7 cm

Page 151

Studies for the Corolised Chair 2012, pen on paper, 42 × 29.7 cm

Research drawing, 2016, pen on paper, 42 × 29.7 cm

My generation of designers has lived through an incredible transition from the hand-drawn, hand-constructed prototypes of the analogue age to the contemporary development of digitally generated and 3-D printed products. This transition has not only changed our perception and awareness of the evolution of objects; it has also, in the process of creation itself, transformed the relationship between the imagination and the result. It clearly marks the evolution of design at the turn of the twenty-first century and with it a radical advance in aesthetics, away from cheap geometries and limited plasticity and toward biomorphic design as well as new, seamless relationships between materials and advanced technologies.

In the 1970s we were hands-on and designers were more akin to sculptors collaborating with model makers. It was a physical process, like that of the handcrafts, with direct levels of trial-and-error design. From these limitations, form masters such as Luigi Colani and Carlo Mollino emerged, designer-engineers with a gift for illustration and a fearless approach to organic, anatomic form. In other words, from the very moment industrial design appeared, the gift to draw and convey reality in one's illustrations was the key to relaying visions of the future to one's audience. These illustrations were incredibly valuable as part of the creative process, even though they were also a universal tool and the public face of creation.

Today, as software programs allow even a basically skilled designer to envisage anything he or she imagines quite effortlessly, the skill to draw by hand has fallen away. Digital renderings have become the new public face, but the ubiquity of such electronic renderings has led to a broad dissemination of images as a style of presentation, such that projects are seen and often dismissed before they are even realized.

For me, however, it has always been important to remain independent of this development in order to preserve a skill that I have and cultivate for myself, personally. This design practice is not only something rare and extremely private; it is in fact central to my sense of artistic exploration and my relationship to art and culture.

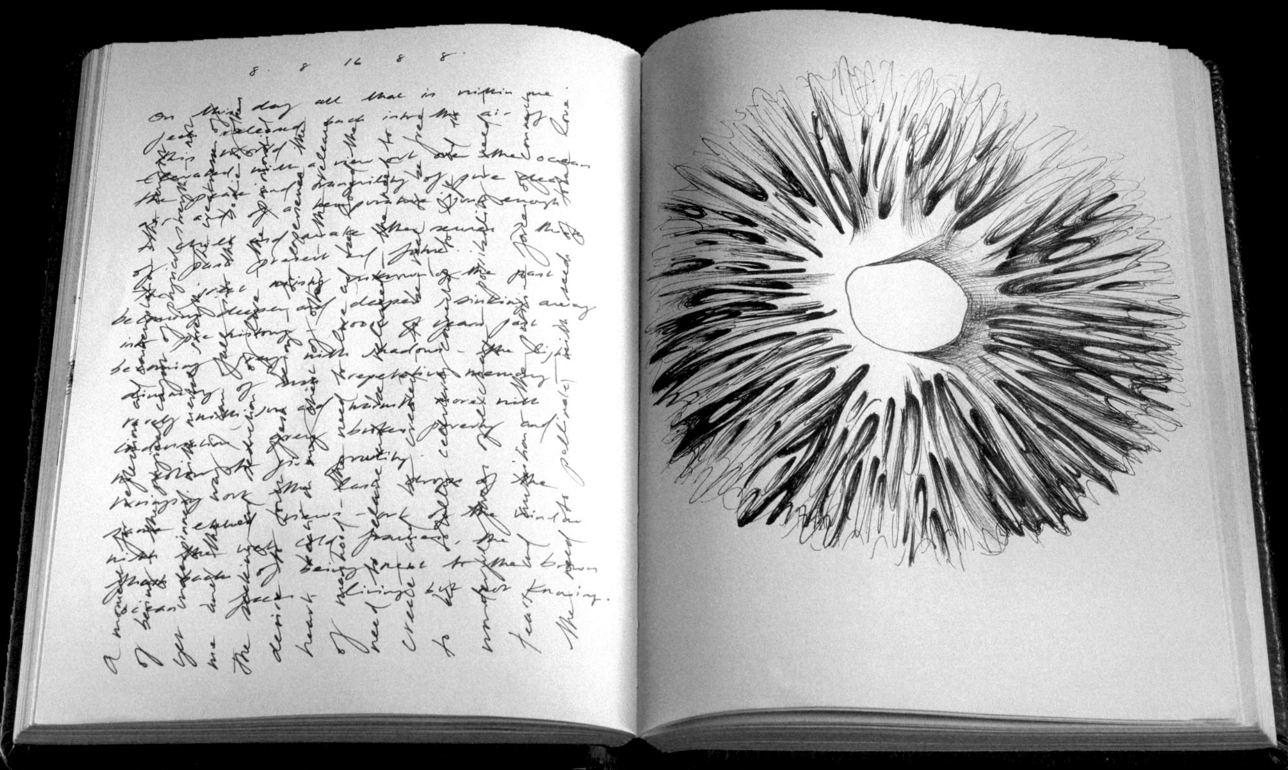

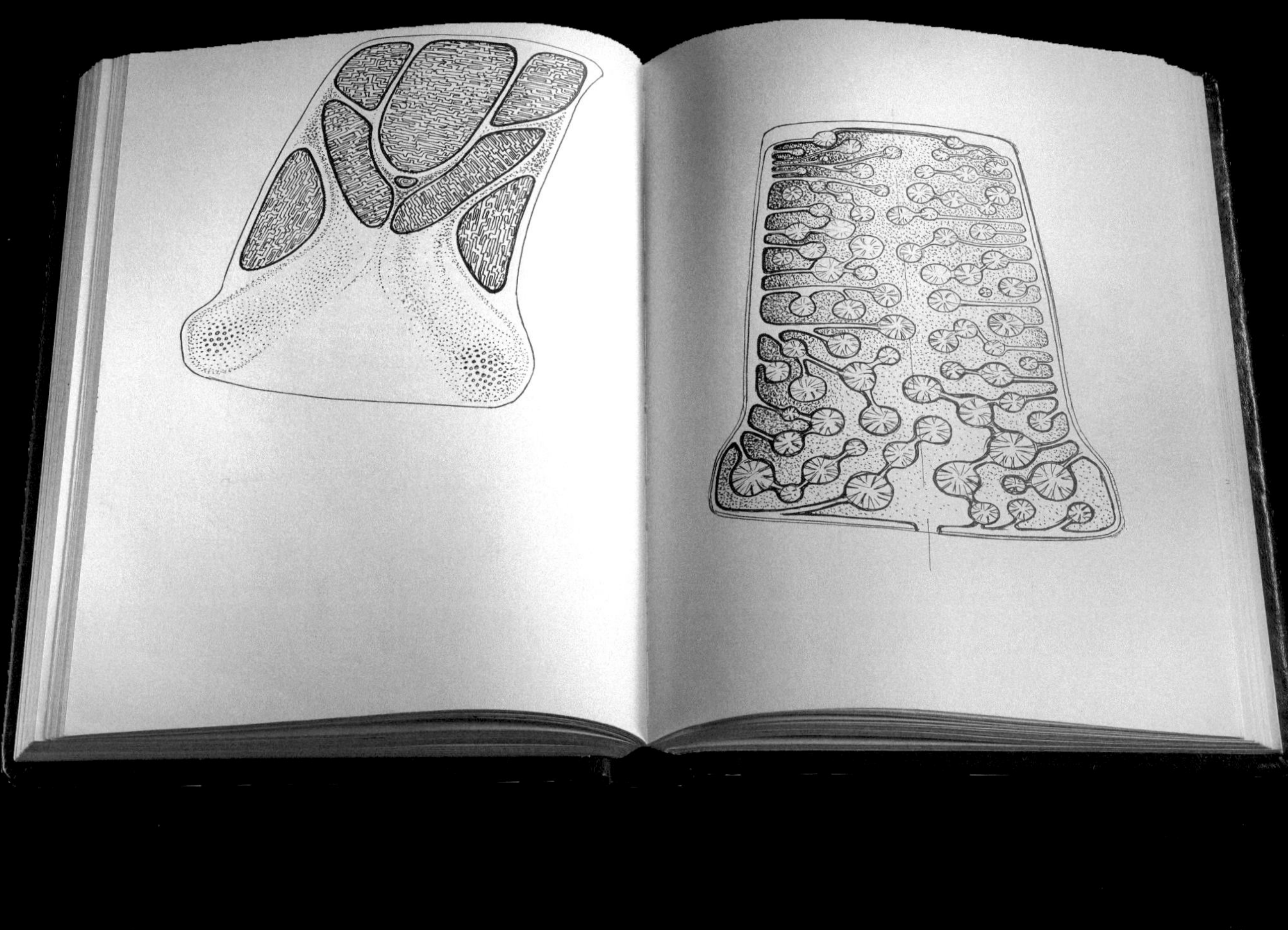

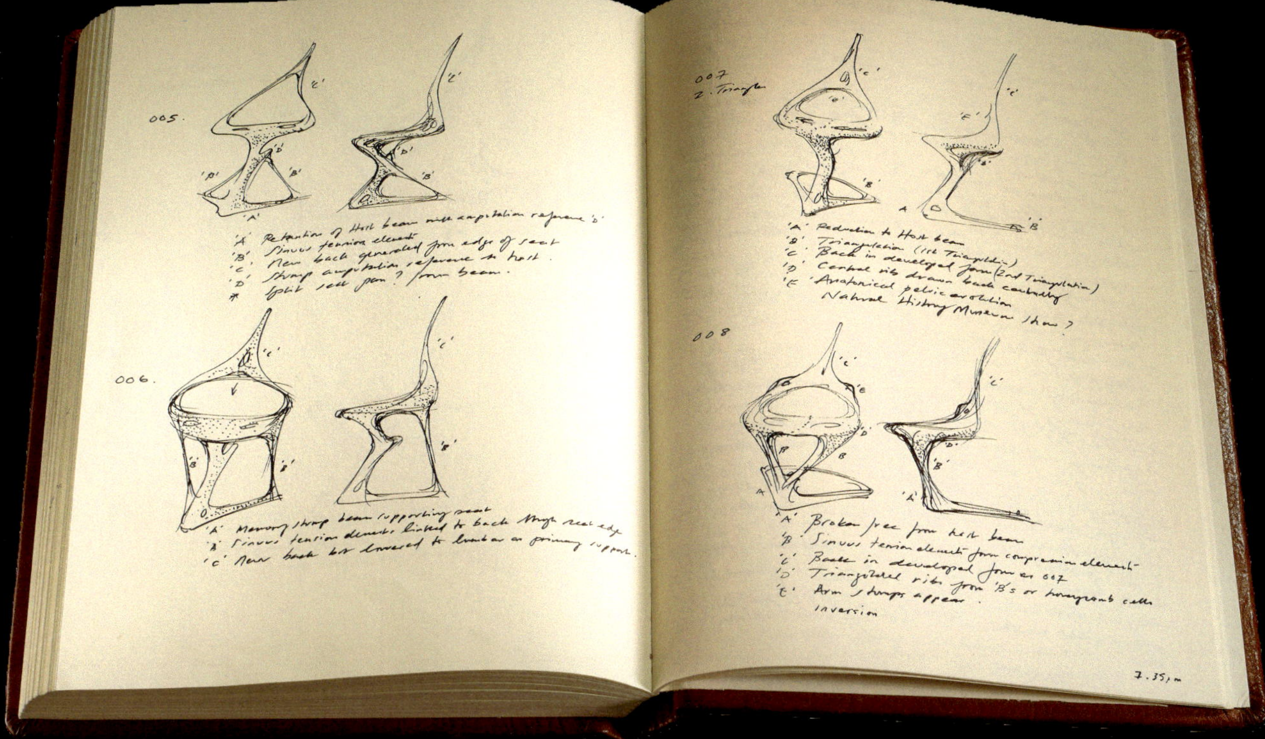
005.
'A' Retention of Host beam with amputation reference 'D'
'B' Sinuous tension element
'C' New back generated from edge of seat
'D' Stump amputation reference to host.
Split seat pan? from beam.
006.
'A' Memory stump beam supporting seat
'B' Sinuous tension elements linked to back thigh seat edge
'C' New back tip lowered to lumbar as primary support.
007
'A' Reduction to Host beam
'B' Triangulation (1st Triangulation)
'C' Back in developed form (2nd Triangulation)
'D' Central rib drawn back centrally
'E' Anatomical pelvic evolution
Natural History Museum Show?
008
'A' Broken free from host beam
'B' Sinuous tension elements form compression element
'C' Back in developed form as 007
'D' Triangulated ribs from 'B's or honeycomb cells
'E' Arm stumps appear.
inversion
7.35pm

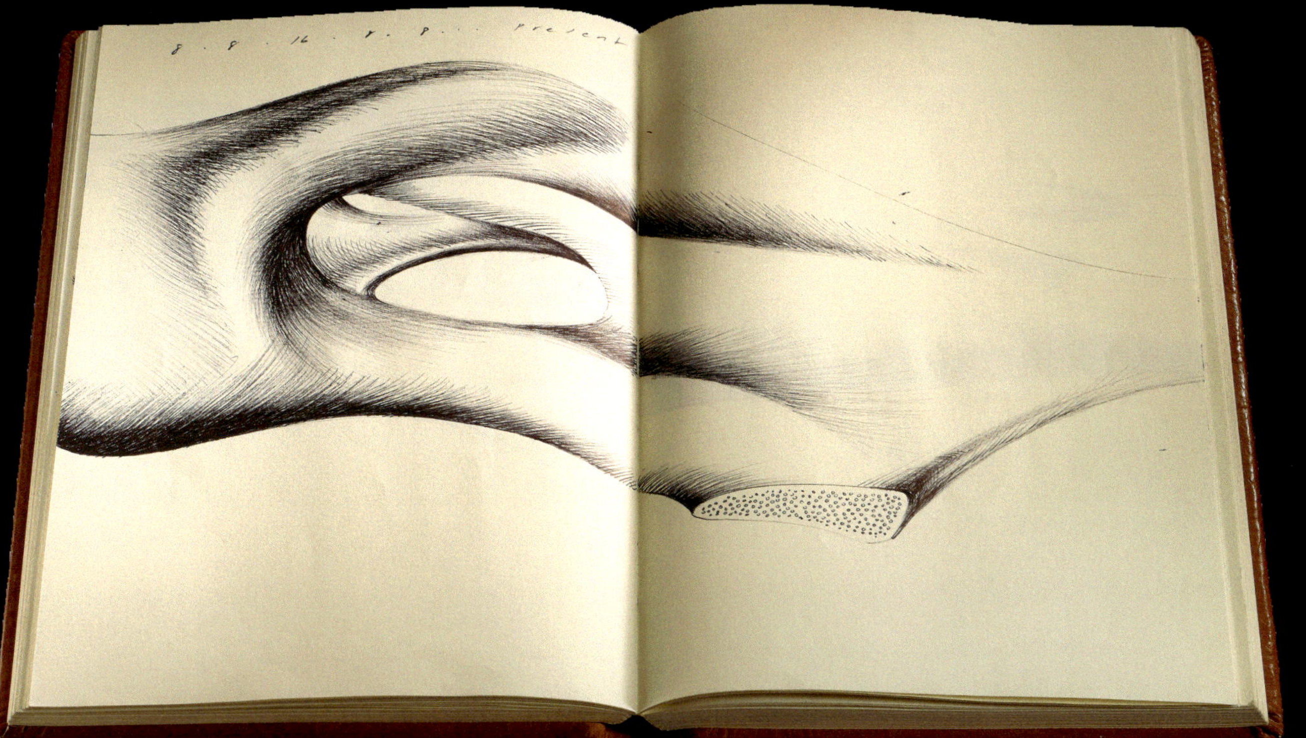
8 . 8 . 16 . T . P . . . present

List of Exhibited Works

Disc Camera 1982–1983
Prototype
Acrylic; 25 × 8 × 3 cm
Manufacturer and collection: Ross Lovegrove
(p. 70)

Eye Digital Camera 1990
Elastomer, rubber; 15 × 8 × 3 cm
Manufacturer: Olympus (Japan)
Model of mold: synthetic resin; various dimensions
Exhibited prototype: silver synthetic resin; 15 × 8 × 3 cm
Collection: Ross Lovegrove
(pp. 74–75)

Solar Seed 1994–1995
Nomadic architecture project
Model: polymers; H.: 75 cm, Dia.: 65 cm
Collection: Ross Lovegrove
(p. 88)

Emma Office System 1994–2000
Unique piece
Corner table: polymer, aluminum, galvanized steel; 75 × 80 × 80 cm
Table: polymer, aluminum, galvanized steel; 70 × 190 × 45 cm
Manufacturer: Herman Miller (United States)
Collection: Ross Lovegrove
(p. 76)

Solar Bud light 1998
Polycarbonate, aluminum, solar-energy LED; H.: 30 cm, Dia.: 15 cm
Manufacturer and collection: Luceplan (Italy)
(p. 89)

Go Chair 1998–2001
Magnesium molded by injection, magnesium powder coating, polypropylene seat; 77.5 × 58.4 × 68.6 cm
Manufacturer: Bernhardt Design (United States)
Collection: Ross Lovegrove
(p. 58)

Airplane seats for JAL 1998–2002
Project for 2,000 seats for the Japan Airlines fleet
Scale model 1:20: mixed materials; 60 × 110 × 60 cm
Manufacturer: Japan Airlines (Japan)
Collection: Ross Lovegrove
(p. 77)

Tŷ Nant mineral water bottle 1999–2001
Polyethylene terephthalate (PET); H.: 23 cm, Dia.: 7 cm
Manufacturer: Tŷ Nant (United Kingdom)
prototype in polythene; H.: 23 cm, Dia.: 7 cm
Collection: Ross Lovegrove
(p. 79)
Illustrations p. 78:
Aluminum molds; 50 × 25 × 25 cm
Computer-generated image
Sketch made on a digital tablet
Collection: Ross Lovegrove

Cetacea lounge chair 2000
Hand-formed copper; 63.5 × 182.2 × 70.5 cm
Edition of 3 and 2 artist's proofs
Manufacturer and collection: Ross Lovegrove
(pp. 54–55)

Megabioform table 2000
Aluminum; 72 × 290 × 162.5 cm
Edition of 6 and 2 artist's proofs
Manufacturer and collection: Ross Lovegrove
(p. 73)

Bamboo Bicycle 2001–2008
Bamboo, aluminum; 102 × 156 × 57 cm
Manufacturer: Biomega (Denmark)
Collection: Ross Lovegrove
(p. 92)

Car on a Stick 2002–2007
Urban ecological transport project
Model: mixed materials; H.: 60 cm, Dia.: 80 cm
Collection: Ross Lovegrove
(pp. 94–95)

Gingko table 2003–2006
Carbon fiber; H.: 72 cm; Dia.: 126 cm
Edition of 12 and 2 artist's proofs
Manufacturer: Ross Lovegrove
Collection: Centre Pompidou, Mnam-CCI
(pp. 62–63)

Biolove bicycle 2003–2011
Carbon fiber and metal; 108 × 185 × 71 cm
Manufacturer: Biomega (Denmark)
Collection: Ross Lovegrove
(p. 65)

Vertebrae 2004
Sculpture: carbon fiber on a digitally milled foam core; 252 × 205 × 63 cm
Edition of 3 and 1 artist's proof
Manufacturer and collection: Ross Lovegrove
(p. 68)

DNA staircase 2005
Combination of fiberglass and unidirectional carbon; H.: 371 cm, Dia.: 231 cm
Edition of 1 and 1 artist's proof
Manufacturer and collection: Ross Lovegrove
(pp. 44, 56–57)

Ridon 2005
Sculpture: carbon fiber on a digitally milled foam rubber core; 105.5 × 181.5 × 40 cm
Edition of 6 and 2 artist's proofs
Manufacturer and collection: Ross Lovegrove
(p. 64)

Swarovski Crystal Aerospace 2005–2006
Prototype of solar-powered car, with the participation of: Sharp Solar Europe, Swarovski Optical Laboratories, Anthony Lo, artistic director of General Motors Europe, Coggiola, Turin
Polycarbonate, aluminum, photovoltaic cells, crystal, silicone; 100 × 420 × 180 cm
Manufacturer and collection: Swarovski UK Ltd (United Kingdom)
(pp. 120–129)

Net Architecture 2005–2006
Architectural project model: 3-D printed; H.: 140 cm, Dia.: 28 cm
Collection: Ross Lovegrove
(p. 105)

Muon speakers 2005–2007
Superformed aluminum; 200 × 38 × 60 cm
Manufacturer and collection: KEF (United Kingdom)
(p. 72)

Supernatural chair 2005–2008
Single shell in polypropylene; 79 × 64 × 56 cm
Manufacturer and collection: Moroso (Italy)
(p. 60)

Hu watch 2005–2008
Titanium; 10 × 5 × 10 cm
Manufacturer: Issey Miyake Watch (Japan)
Collection: Ross Lovegrove
(p. 71)

GT110 suitcase 2006–2007
Carbon fiber, Kevlar; 65 × 80 × 28 cm
Manufacturer: Globe-Trotter (United Kingdom)
Collection: Ross Lovegrove
(p. 65)

Generator House 2006–2008
Project for a self-sufficient house using solar energy
Computer-generated images
Exhibited work: Stereolithograph (SLA) 3-D printed on acrylic tubes; H.: 200 cm, Dia.: 115 cm
Production and collection: Swarovski UK Ltd (United Kingdom)
(pp. 90–91)

Andromeda light fixture 2006–2008
Suspended ceiling lamp: steel with painted finish, hard polyurethane foam, mirrors, LED; 27 × 80 × 46 cm
Manufacturer: Yamagiwa (Japan)
Collection: Ross Lovegrove
(p. 61)

Solar Tree lamp 2006–2010
Steel, concrete, LED, photovoltaic cells; H.: 550 cm, Dia.: 441 cm
Manufacturer: Artemide (Italy) and Ross Lovegrove
Collection: Ross Lovegrove
(pp. 80, 86–87)

Liquid Carbon bench 2007
Carbon fiber on a digitally milled polyurethane foam core; 60 × 285 × 85 cm
Edition of 10 and 2 artist's proofs
Manufacturer and collection: Ross Lovegrove
(p. 67)

Calgat [Can a Lily Grow a Telephone] 2008
Computer-generated image
Exhibited model (cf. detail p. 82): 3-D printed; 80 × 80 × 80 cm
H.: 47 cm, Dia.: 80 cm
Collection: Ross Lovegrove
(p. 93)

Cellular Automata 2008
Research group printed in 3-D
5 unique pieces
Stereolithograph; 60 × 45 × 45 cm
Computer-generated images
Collection: Ross Lovegrove
(pp. 102–103)

Novartis staircase 2009
Kevlar
Manufacturer: Novartis Pharma AG (Switzerland)
Novartis Campus, Basel
Architect: David Chipperfield
Exhibited model: 3-D printed; 65 × 52 × 17 cm
Collection: Ross Lovegrove
(p. 69)

Cosmic Angel light 2009–2011
Ceiling light: aluminum, polymer; 25 × 178 × 80 cm
Manufacturer and collection: Artemide (Italy)
(p. 108)

Kenzo perfume bottle 2010–2011
Modelization studies
Drawing by Ross Lovegrove and glass prototype; 13 × 9 × 3 cm
Manufacturer: Kenzo (Japan)
Collection: Ross Lovegrove
(pp. 116–117)

Liquid Shelves 2011
Aluminum; 185 × 300 × 50 cm
Edition of 4 and 2 artist's proofs
Manufacturer and collection: Ross Lovegrove
(p. 73)

Corolised Chair 2012
Alumide printed in 3-D; 70 × 45 × 45 cm
Edition and collection: Ross Lovegrove
Edition of 100
(pp. 112–113)

Instinctive Override 2012
Installation: video projection on fiberglass form; 126 × 406 × 167 cm
Collection: Ross Lovegrove
(p. 147)

New Nature floor lamp 2012
Steel, polymer; H.: 188 cm, Dia.: 28 cm (base)
Manufacturer and collection: Artemide (Italy)
(p. 59)

Nebula wall lighting piece 2012
Aluminum, opaline methacrylate, polished chromed metal, LED.; Dia.: 80 cm, D.: 14 cm
Manufacturer and collection: Artemide (Italy)
(p. 104)

Moot chair 2012–2013
Carbon fiber; 80 × 59 × 46 cm
Edition of 58 and 2 artist's proofs
Manufacturer: Established & Sons (United Kingdom)
Collection: Ross Lovegrove
(p. 66)

Lasvit LiquidKristal pavilion 2012
LiquidKristal glass, steel framework, Barrisol stretched ceiling; 400 × 1200 × 800 cm
Manufacturer and collection: Lasvit (Czech Republic)
(pp. 140–143)

Twin'Z electric car 2012–2013
Mixed materials; 152 × 363 × 170 cm
Manufacturer: Renault (France)
Collection: Renault Modern Art Collection, Boulogne-Billancourt
(pp. 130–139)

Biophilia chair 2014
Rotomolded polyurethane; 103 × 47 × 56 cm
Manufacturer and collection: Vondom (Spain)
(p. 115)

Diatom chair 2014
Natural molded aluminum varnished with a glossy, transparent protective coating; 77 × 72 × 63 cm
Manufacturer and collection: Moroso (Italy)
(pp. 110–111)

Ilabo shoes 2015
3-D printed nylon and polyurethane; 35 × 30 × 12 cm
Edition of 50 and 1 artist's proof
Manufacturer: United Nude (Netherlands)
Collection: Ross Lovegrove
(pp. 96, 106–107)

Chlorophilia 2 suspension light 2015
Ceiling light: polycarbonate, aluminum, LED; H. max.: 200 cm, Dia.: 77.5 cm
Manufacturer and collection: Artemide (Italy)
(p. 109)

Barrisol Cocoon 2016–2017
Sheets of Barrisol Mylar on an aluminum structure; 265 × 840 × 840 cm
Computer-generated images
Manufacturer and collection: Barrisol Normalu (France)
(pp. 144–146)

Spine Chair 2017
Prototype
3-D print; 70 × 45 × 45 cm
Manufacturer and collection: Moroso (Italy)
(p. 114)

Biography

Born 1958 in Cardiff, Wales
Lives and works in London

1980
Receives bachelor of arts cum laude in industrial design from Manchester Polytechnic.

1983
Receives master of design from the Royal College of Art.
At the outset of the 1980s works at Frog Design in West Germany on projects such as the Walkman for Sony and computers for Apple Computers.

1984–1987
Works as a consultant for Knoll International in Paris, where he designed the Alessandri Office System.
Invited to join the Atelier de Nîmes along with Jean Nouvel and Philippe Starck; works as a consultant for Cacherel, Louis Vuitton, Hermes, and Dupont, among others.

1986
Returns to London, where he pursues projects for Kartell, Cappellini, Moroso, Luceplan, Driade, Vitra, Herman Miller, Artemide, Aibus Industries, Japan Airlines, and Yamagiwa Corporation, as well as others.

1990
Opens Studio X in Notting Hill. Realizes the chair Supernatural.

1993
Curates the first exhibition of the permanent collection at the London Design Museum.

1994
Begins work on projects using solar energy.

2005
Develops the Swarovski Crystal Aerospace concept car.

2012
Starts collaborating with Renault on the Twin'Z electric car.

2014
Exhibits the Barrisol Pavilion at the 24th Biennale Interieur, Kortrijk, Belgium.

Awards and Distinctions

1998
Médaille de la Ville de Paris
George Nelson Award

1999
iF Industry Forum Design Award, Hannover

2000
Good Design Award, conferred by *ID* magazine

2001
Designer of the Year, awarded by the magazine *Architektur & Wohnen,* Hamburg

2004
Final selection for the Prince Philip Design Prize
Prix Janus, Paris
Royal Designer for Industry Award, conferred by the Royal Society of Arts, London

2005
World Technology Award, given by *Time* magazine and CNN

2006
Wall Paper Design Award

2007
Condé Nast Vogue Traveller Ecology Prize for the Solar Tree, MAK, Vienna

2009
Batimat Design Award

2010
Red Dot Award: Best of the Best

2015
Elle Déco International Design Awards for the Diatom chair

2016
German Design Award for the Wanderlust collection from Kiko

Collections

Works by Ross Lovegrove are found in the permanent collections of many museums around the world, notably the Museum of Modern Art (MOMA) in New York, the Design Museum in London, the Vitra Design Museum in Weil am Rhein (near Basel), and the Musée national d'art modern at the Centre Pompidou in Paris.

List of Exhibitions

Ross Lovegrove: Objects, Volvo Centre, Stockholm, 1997.

Ross Lovegrove: Design, Danish Museum of Decorative Art, Copenhagen, February 1998.

Organic Dreams, IDEE Gallery, Tokyo, December 1999.

Sensual Organic Design, Yamagiwa Corporation, Tokyo, October 2000.

Material Transition, Rheinauen Space, Cologne, in conjunction with the Designer of the Year Award, *Architektur & Wohnen* magazine, January 2001.

Expanding the Gap, with Greg Lynn and Tokujin Yoshioka, Rendel & Spitz gallery, Cologne, January 14–20, 2002.

Delighted by Corian, with Marc Newson and James Irvine, Milan Furniture Fair, Milan, April 14–17, 2003.

Latent Utopias, Steirischer Herbst Fair, Graz, November 30, 2003.

Designosaurs, in conjunction with the exhibition *From Archaeology to Architecture*, Segheria Gallery, Milan, April 2004.

Superliquidity, Le Bain gallery, Tokyo, March 2005.

Endurance: Limited Edition Art Pieces, Phillips de Pury & Company, New York, March 21–April 4, 2007.

Exhibition of Solar Tree and Swarovski Crystal Aerospace, MAK, Vienna, October 2008.

Cellular Automation—Origin of Species 2, in conjunction with the exhibition *Second Nature*, 21_21 Design Sight, Tokyo, October 17, 2007–January 18, 2008.

Design Cities, Istanbul Modern, Istanbul, April 23–August 10, 2008.

Liquid Space, "Swarovski Crystal Palace" stand, Art Basel, Miami, November 30–December 4, 2008.

Primordial, Galerist gallery, Istanbul, January 24–February 28, 2009.

Freiburg Solar Show, Freiburg im Breisgau, 2009.

Endless, Galleria Cardi Black Box, Milan, September 12–October 1, 2011.

Lasvit LiquidKristal pavilion, Milan Furniture Fair, Milan, April 4–9, 2012.

Sharp Solar and Artemide stand, Clerkenwell Design Week, London, May 22–24, 2012.

23rd Biennale Interieur, Kortrijk, October 18–20, 2012.

Vondom stand, Milan Furniture Fair, Milan, April 9–14, 2013.

Naturaliser l'architecture, ArchiLab, FRAC Centre, Orléans, September 14, 2013–March 30, 2014.

London Design Festival, London, September 15–19, 2014.

24th Biennale Interieur, Kortrijk, October 17–26, 2014.

Moooi Carpets stand, Milan Furniture Fair, Milan, April 14–19, 2015.

Moroso stand, Clerkenwell Design Week, London, May 19–21, 2015.

United Nude stand, New York Design Week, New York, May 8–June 3, 2015.

London Design Festival, London, September 19–27, 2015.

In-Formation, for Alcantara®, IMM, Cologne, January 16–22, 2017.

Selected Bibliography

Monograph

• Lovegrove, Ross, and Paola Antonelli. *Supernatural: The Work of Ross Lovegrove.* New York, 2004.

General Publications

• Benyus, Janine M. *Biomimicry: Innovation Inspired by Nature.* New York, 2002.
• Beukers, Adriaan, and Ed Van Hinte. *Lightness.* Rotterdam, 1998.
• Fiell, Charlotte, and Peter Fiell. *Designing the Twenty-First Century.* Cologne, 2001.
• Griffiths, Alyn. *21st Century Lighting Design.* London, 2014.
• Lovegrove, Ross. *International Design Yearbook* 17. London, 2002.
• Lupton, Ellen. *Skin: Surface, Substance + Design.* New York, 2002.
• Papadakis, Alexandra, and Andreas Papadakis. *Innovation: From Experimentation to Realization.* London and New York, 2003.
• Rendel, Martin, and René Spitz: *Expanding the Gap / Das Weite suchen.* London and Stuttgart, 2002.

Exhibition Catalogues

• *Ceramic Network 1999: Earth & Water: Contemporary Ceramic Potential in Architecture: Ross Lovegrove.* Edited by Jeanne Quéheillard et al. Exh. cat. Centre de recherche des arts du feu et de la terre. Limoges, 1999.
• *Charles Eames Furniture from the Design Collection.* Edited by Arthur Drexler and Eliot Noyes. Exh. cat. Museum of Modern Art. New York, 1973.
• *Impression 3D l'usine du future: 70 créations innovantes.* Edited by François Brument, Maëlle Campagnoli. Exh. cat. Le lieu du design. Paris, 2016.
• *Latent Utopias: Experiments within Contemporary Architecture.* Edited by Zaha Hadid and Patrik Schumacher. Exh. cat. Steirischer Herbst 2002, Landesmuseum Joanneum, Graz. Vienna, 2002.
• *Lighting Affinities.* Edited by Vanni Pasca. Exh. cat. Luce Plan Factory Gallery. Milan, 1996.
• *Mutant Materials in Contemporary Design.* Edited by Paola Antonelli. Exh. cat. Museum of Modern Art. New York, 1995.
• *Nature Design: From Inspiration to Innovation.* Edited by Angeli Sachs. Exh. cat. Museum für Gestaltung Zürich. Zurich, 2007.
• *Objects of Design from the Museum of Modern Art.* Edited by Paola Antonelli. Exh. cat. Museum of Modern Art. New York, 2003.
• *Ross Lovegrove: Objects.* Edited by Stefan Ytterborn and Peter M. Fiell. Exh. cat. danske Kunstindustrimuseum, Copenhagen. Stockholm, 1997.

Articles

• Agboton, Guy-Claude. "Ross Lovegrove, l'homme qui venait d'ailleurs." *IDEAT* (June 2009), pp. 114–117.
• "All You Need Is Lovegrove." *The Sunday Times Magazine* (March 31, 1996).
• Benyus, Janine M. "Organic Design: Ross Lovegrove." *Ppaper* no. 52 (September 2007), pp. 25–56.
• Bergamini, Alessandra. "Composing through Dots and Lines: Poetry in Light." *Ottagono* no. 261 (June 2013), pp. 112–117.
• Bettens, Walter, and Ross Lovegrove. "When Love Comes to Town." *DAMN* (October 2007), pp. 54–61.
• Billaud, Jean-Pascal. "Lovegrove and Brown." *City* no. 26 (September 1986).
• Braunstein-Kriegel, Chloé. "Ross Lovegrove: Harmonies organiques." *Beaux-Arts Magazine* (2009), pp. 146–147.
• Brigi, Elena. "Luceplan tra le migliori interpreti del design." *Ottagono* no. 219 (April 2009), pp. 92–95.
• "British Grace." *Form* no. 1 (1995).
• Bruno, Giulia. "From Marketing to Nature." *DDN Design Diffusion News* no. 169 (June 2010), pp. 186–201.
• Bullivant, Lucy. "Ross Lovegrove designer." *Intramuros* no. 56 (December 1994–January 1995).
• Capella, Julie. "In/Off, Salone del Mobile." *Domus* no. 784 (July 1996).
• Cattermole, Pierluigi. "Senseware—Tokyo Fiber." *Experimenta* 64, no. 1 (November 2003), p. 54.
• "Clandestine Cool." *Marie Claire Japan* no. 150 (1994).
• "Commentaire sur le vif." *Architecture Intérieure* no. 205 (May–June 1985).
• "Connolly luxe calme et volupté." *Intramuros* no. 59 (June 1995).
• "The Conran Foundation." *Abitare* no. 324 (December 1993).
• "Design DNA." *Clear* (December 2006–January 2007), pp. 60–64.
• "Design from London." *Architecture d'Aujourd'hui* no. 250 (April 1987).
• "Design High." *Wohn Design* (November–December 1995).
• "Design: The Best and Worst of 2001." *Time Magazine* (December 24, 2001).
• "A Designer with Bottle." *The Independent* (May 18, 2002).
• "English Leather." *ID Magazine* (May–June 1996).
• Estrada, Maria Helena. "Ross Lovegrove Etica e Estética." *Arc Design* no. 27, São Paulo (September–October 2002), pp. 20–22.
• Ezersky, Ivan. "Ross Lovegrove: Designer of the Year." *NG-Koulisa,* Moscow (February 2, 2001).
• Fiell, Peter. "La nuova natura." *Domus* no. 795 (July 1997).
• "The 50 Best Ways to Light up Your Life." *The Independent* (October 20, 2001).
• Fitoussi, Brigitte. "Ross Lovegrove: vers un minimalisme organique." *L'Architecture d'aujourd'hui* no. 312 (September 1997), pp. 102–113.
• "5.000 Türgriffe in einem Gebäude . . ." *Möbel Raum Design* (October–November 1995).
• "A Good Hiding." *The Magazine* (April 9, 1995).
• Hamaide, Chantal. "Renault joue l'esprit Cartoon." *Intramuros* no. 166 (May–June 2013), pp. 76–79.
• "I pensieri della amno di Ross Lovegrove." *Interni* no. 444 (October 1994).
• "In nuova forma." *Casa Vogue* (September 24, 1992).
• "Japan Airlines to Use Design by Ross Lovegrove in First-Class Cabin." *Axis* no. 94, Tokyo (November–December 2001), pp. 12–14.
• Jow, Tiffany. "Endurance Test." *Surface* no. 66 (Summer 2007), p. 32.
• Kohuis, Merel. "Ross Lovegrove: Talk." *Frame* (March 25, 2009), pp. 16–21.
• Koller, Gerhard. "Solar Tree Vienna, Austria." *Bright Book* (March 2008), pp. 8–15.
• "Liquid Solids." *Esquire* (March 2000).
• "Lovegrove." *Design Report* no. 10 (October 1995).
• Lovegrove, Ross. "Cover Interview." *Axis* no. 97, Tokyo (May–June 2002), pp. 1–2.

• ———. "Design and Lightnes." *Domus* no. 982 (July–August 2014), pp. 90–97.
• ———. "Progretti Project." *Domus* no. 800 (January 1998), pp. 78–82.
• ———. "Super-Natural: Parametricism in Product Design." *AD. Architectural Design* 86, no. 2 (March–April 2016), pp. 102–107.
• ———. "You're not going to wash your socks in my sink." *Frame* (March–April 2006), pp. 26–27.
• Lovegrove, Ross, and Franco Raggi. "Light Tech." *Flare Magazine* (2004), pp. 92–96.
• Lovegrove, Ross, and IDEAT. "Le studio organique du designer Ross Lovegrove." *IDEAT* no. 90 (February 2012), pp. 138–143.
• "Métropole." *Intramuros* no. 74 (December 1997), pp. 50–53.
• Morteo, Enrico. "La qualita della scrittura." *Domus* no. 745 (January 1993).
• "Next Big Thing: Ross Lovegrove's New Series for Artemide Is Just the Latest in the Company's 50-Year Investment in Design." *Metropolis* (May 2009), pp. 103–110.
• "Una nuova fiducia scultorea." *Domus* no. 758 (March 1994).
• "Old England Discovers Design." *Blueprint* no. 117 (May 1995).
• Pasca, Vanni. "Ross Lovegrove: Un langage organique et sensuel." *Driade Edizioni* no. 4 (April 1997), pp. 37–63.
• Picchi, Franchesca. "Organic Minimalism." *Domus* no. 838 (June 2001).
• Pragnya, Rao. "Conversations with Ross Lovegrove in Dubai." *Elle Décor* (October–November 2009), pp. 75–76.
• Rappolt, Mark. "Go with the Flow: Ross Lovegrove's Adventures in Fluid Form." *Modern Painters* (April 2006), pp. 40–43.
• Romnalli, Marco. "Figure of 8 chair." *Domus* no. 743 (November 1992).
• "Ross Lovegrove et Julian Brown." *Intramuros* no. 19 (July–August 1988).
• "Ross Lovegrove." *Galleries Magazine* (February–March 1994).
• "7 questions à Ross Lovegrove, le gourou de la courbe." *Harmonies* (July–August 2016), pp. 122–124.
• Sommariva, Elena. "Twin'Z concept car: In Lovegrove with Renault." *Domus* no. 969 (May 2013).
• "Spaazi High-touch." *Interni* no. 445 (November 1994).
• Starr, Nina. "Design Matters." *T Qatar* (September–October 2016), pp. 46–50.
• "Uno studio protipo." *Domus* no. 862 (September 2003).
• "Les techno-inventifs: True originators." *AD Collector* no. 7 (September 2012), pp. 20–27.
• Thompson, Henrietta. "The Return of Captain Organic." *The Telegraph* (December 19, 2013).
• Tommasini, Maria Cristina. "Light Design." *Domus* no. 925 (May 2009,) pp. 89–95.
• ———. "Organic Essentialism." *Domus* (January 2008), pp. 100–104.
• Ubeda, Ramon. "La Aventura Española de Ross Lovegrove." *Diseño interior* no. 122 (2002), pp. 60–62.
• Wilson, Michael. "Latente Utopien/Enactments of the Self." *Artforum International* (February 2003), pp. 11–13.
• Woodcock, Ben. "Ross Lovegrove a Step Ahead." *Pol Oxygen* no. 7 (2003), pp. 43–47.
• Zagula, Artur. "Ross Lovegrove, Cooper Union." *Design_pl* no. 1 (2011), pp. 3–16.

Online Articles

• Azzarello, Nina. "United nude + 3D systems present re-inventing shoes during Milan design week." designboom.com (April 14, 2015).
• Chin, Andra. "Barrisol pavilion + lighting by Ross Lovegrove at biennale interieur." designboom.com (October 20, 2014).
• ———. "Ross Lovegrove creates infinitely stackable Diatom chair for Moroso." designboom.com (April 9, 2014).
• Fairs, Markus. "Ross Lovegrove for KEF." dezeen.com (March 28, 2007).
• "The Gingko Carbon Table." centrepompidou.fr., n.d.
• Hermann, Christoph. "Infinity Loop large light objects by Studio Ross Lovegrove for Barrisol." christoph-hermann.com (April 2014).
• Hobson, Ben. "Moroso 'gave me a chance' says designer Ross Lovegrove." dezeen.com (November 7, 2014).
• ———. "Watch Ross Lovegrove's keynote presentation live from inside 2016." dezeen.com (November 18, 2016).
• Howarth, Dan. "Zaha Hadid, Ben van Berkel and more design 3D printed shoes for United Nude." dezeen.com (April 13, 2015).
• "Ilabo—Ross Lovegrove for United Nude." arturotedeschi.com (April 2015).
• Jarcy, Xavier de. "Ross Lovegrove: organic designer "à Notting Hill." telerama.fr (April 11, 2009).
• Kwok, Natasha. "A look inside Ross Lovegrove's four-storey home and studio in London." designboom.com (October 21, 2015).
• Lovegrove, Ross. "The Gingko Carbon Table by Ross Lovegrove." phaidon.com (July 23, 2010).
• Mills, Joe. "LDN by Ross Lovegrove and NYC by KiBiSi for Biomega." dezeen.com (August 20, 2010).
• Mull, Olivia. "Ross Lovegrove creates chair for Moroso using metal-pressing technology for cars." dezeen.com (April 18, 2014).
• "Re-Inventing Shoes." unitednude.com (April 2015).
• "Ross Lovegrove: Nebula + Botanic Rain for Artemide." designboom.com (April 17, 2012).
• "Tokyo Design Week 08: 'Cellular Automation' by Ross Lovegrove." designboom.com (November 13, 2008).
• Tucker, Emma. "Ross Lovegrove downsizes giant speaker to create wireless Muo device." dezeen.com (October 8, 2015).

Conference Presentations by Ross Lovegrove

• "Convergences." Centre Pompidou, Paris, 2015.
• "The Future of Organic Design." DLD09, London, 2009.
• "Global Strategy." Innoprom, Ekaterinburg, August 2013.
• "Organic Design, Inspired by Nature." TED Talk, New York, 2005.
• "Out of this World." London, September 15, 2015.
• "Revealing Energy." TED Global 2009, session 9, Oxford, UK, July 23, 2009.
• "Transformed by Light." Arena Berlin, Berlin, November 18, 2016.
• "2015 Be3Dimensional Design Conference." Winter Garden Theatre, Toronto, October 23–24, 2015.

Exhibition

Curator
Marie-Ange Brayer

Deputy curator
Olivier Zeitoun

Exhibition Manager
Marie-Annick Hamon

Exhibition Design
Ross Lovegrove
with **Stefan Bassing** and
Charline Ronzon-Jaricot

Architect
Laurence Le Bris

Artworks Registrar
Marie-Colleen Payen

Management of Exhibition Space
Anne-Marie Spiroux

Lighting Design
Arnaud Jung

Graphic Design
Bastien Morin

Cultural Mediation
Stéphanie Chaillou

MUSÉE NATIONAL D'ART MODERNE – CENTRE DE CRÉATION INDUSTRIELLE

Director
Bernard Blistène

Deputy Directors
Brigitte Leal (Collections)
Catherine David (Research and Globalization)
Frédéric Migayrou (Industrial Design)
Didier Ottinger (Cultural Programming)

Chief Administrative Officer
Xavier Bredin

Bibliothèque Kandinsky
Didier Schulmann
Curator and Head of Department

DEPARTMENT OF CULTURAL DEVELOPMENT

Director
Kathryn Weir

Head of the Spoken Word Unit
Jean-Max Colard

Programming Manager
Romain Lacroix

Production
Martial Lhuillery

Publication

Editor
Marie-Ange Brayer
assisted by
Olivier Zeitoun

Subeditor
Geneviève Munier

Research
Thomas Bertail
Marie Siguier
Mathilde Vallée

Drafting of Entries
Marie Siguier (MS)
Mathilde Vallée (MV)

Graphic Design and Layout
Laure Cérini

Image and Rights Management
Clarisse Deubel

PUBLICATIONS DEPARTMENT

Director
Nicolas Roche

Head of Publishing
Claire de Cointet

Editorial Manager
Françoise Marquet

Head of Sales
Élise Albenque

Sales Officer
Francesca Baldi

Receipts and Contracts Manager
Mathias Battestini

Artistic Credits

For all the reproduced works by Ross Lovegrove:
© Ross Lovegrove

© Adagp, Paris 2017 for the works by
Tony Cragg and Anish Kapoor
© Frei Otto + Partners
Building of the Centre Pompidou:
© Studio Piano & Rogers, architects
© Christina Brodie, 2017
Luigi Colani: © Colani Design/Colani Trading AG
Charles and Ray Eames: © 2017 Eames Office, LLC
(eamesoffice.com)
© Antony Gormley
Carlo Mollino: Museo Casa Mollino, Turin
Henry Moore: Reproduced by permission of
The Henry Moore Foundation
© Bruno Munari, 1949 – All rights reserved
to Maurizio Corraini s.r.l.
© NOX/Lars Spuybroek
© Tokujin Yoshioka
© Zaha Hadid Architects

Photo Credits

© Artemide: pp. 104, 108, 109
© A. Barrington Brown/SPL – Science Photo Library/
Biosphoto: p. 22
Photo John Berens: pp. 35 bottom, 98 bottom right
© Anthony Bernier: pp. 133, 136, 138 top, 139
© Christina Brodie, 2017: p. 99 center right
© Luis Casiano / Biosphoto: p. 18 bottom
© Centre Pompidou, MNAM-CCI/Jacques Faujour/
Dist. RMN-GP: p. 21 top
© Centre Pompidou, MNAM-CCI/Philippe Migeat/
Dist. RMN-GP: p. 37
© Colani Design/Colani Trading AG-photo Tom Vack:
p. 83 center right
© Simona Cupoli: pp. 43, 140-141 (Courtesy of Lasvit),
143 bottom (Courtesy of Lasvit)
© Michele d'Ottavio: p. 123
© Eames Office, LLC: p. 33 top
Photo ESO/VISTA/J. Emerson. Acknowledgment:
Cambridge Astronomical Survey Unit: p. 83 top left
© Gianpiero Ferrari/FLPA – Frank Lane Picture Agency/
Biosphoto: p. 46 bottom left
© Pascal Goetgheluck/Biosphoto: pp. 47 center left,
82 bottom left
© Steve Gschmeissner/SPL – Science Photo Library/
Biosphoto: p. 17
© Ken Hickman: p. 98 top left
Photo David Higginbotham/NASA: p. 83 bottom left
© Institut für Leichtbau Entwerfen und Konstruieren,
Universität Stuttgart: p. 28
© Pierre Jusselme: pp. 125, 126 top, 127 top
Courtesy of Knoll Inc.: p. 19
© Scott Leslie/Minden Pictures/Biosphoto: p. 99 bottom
right
Photo Library of Congress: pp. 26
© Ross Lovegrove Studio: pp. 46 center left, 47 top right
and center right, 56, 57 bottom right, 58, 59, 61, 62, 68,
69 bottom left, 73, 74, 78 center and right, 80, 82 top
right, center left and bottom right, 83 top right, center
left and bottom right, 86, 88, 90, 91, 93-95, 98 top right,
center right and bottom left, 99 top right and bottom
left, 103, 105, 107 top left and bottom left, 112, 114,
116–118, 124 top, 126 bottom, 127 bottom, 128, 129, 134,
135, 137, 138 bottom, 143 top right, 143 top left (courtesy of
Lasvit), 145–148, 150, 151
© Greg Lynn FORM: p. 30
© Chris Mellor/Getty Images: p. 47 bottom right
© Pat Morris/Ardea/Biosphoto: p. 18 top
© Museo Casa Mollino, Turin: p. 21 bottom
© Nacása & Partners Inc.: p. 32
© Pedro Narra/naturepl.fr: p. 46 center right
Photo NASA: p. 82 top left
Courtesy NASA/JPL–Caltech: p. 23
© Novartis: p. 69 top left, top right and bottom right
© NOX/Lars Spuybroek: p. 31
© Alessandro Paderni: pp. 99 center left, 111
© Bert Pijs/NiS/Minden Pictures/Biosphoto: p. 99 top
left
© John Ross Studio: cover and cover sheets, endpapers,
pp. 44, 46 top right, 47 bottom left, 54, 55, 57 top left,
top right and bottom left, 60, 63–67, 70–72, 75–77, 78
left, 79, 89, 92, 102, 113, 120–121, 124 bottom, 130–131
Royal Collection Trust © Her Majesty Queen
Elizabeth II, 2017/Bridgeman Images: p. 47 top left
Service Commun de la Documentation de l'Université
de Strasbourg: p. 25
© Jonathan Singer: p. 82 center right
© Andrew Syred/SPL – Science Photo Library/
Biosphoto: p. 98 center left
© Tate, London, 2017: pp. 34 bottom, 46 top left
Reproduced by permission of The Henry Moore
Foundation: pp. 20 bottom, 34 top
© Arturo Tedeschi for Ross Lovegrove Studio: p. 96
© 2017. Digital image, The Museum of Modern Art,
New York/Scala, Florence: p. 33 top
© United Nude: p. 107 right
© Vondom: pp. 46 bottom right, 115
Photograph by Stephen White, London: p. 35 top
© Dirk Wiersma/SPL – Science Photo Library/
Biosphoto: p. 20 top
© Zaha Hadid Architects: p. 29

First published in France in 2017
by Centre Pompidou, Paris

French title:
Ross Lovegrove – Convergence

Typesetting and Production
Sieveking Verlag, Munich

Translations
Michael Abbot (French–English):
pp. 4 – 5, 8 – 9, 45 – 49, 54 – 121, 130 – 148
Gila Walker (French–English):
pp. 16 – 23, 24 – 42

Copyeditor
Joann Skrypzak

Published by
Sieveking Verlag, Munich, 2017

ISBN 978-3-944874-76-0

Printed in Germany

Cover:
Ross Lovegrove
Eye Digital Camera, 1990
Elastomer, rubber; 15 × 8 × 3 cm
Manufacturer: Olympus (Japan)
Ross Lovegrove Collection

Endpapers:
Interior of Ross Lovegrove's studio,
Notting Hill, London, 2015

Photoengraving:
Arciel Graphic, Paris

Centre Pompidou

Published in conjunction with the exhibition presented
at the Centre Pompidou, Galerie 3, from April 2 to July 3, 2017,
as part of the event *Mutations/Créations*

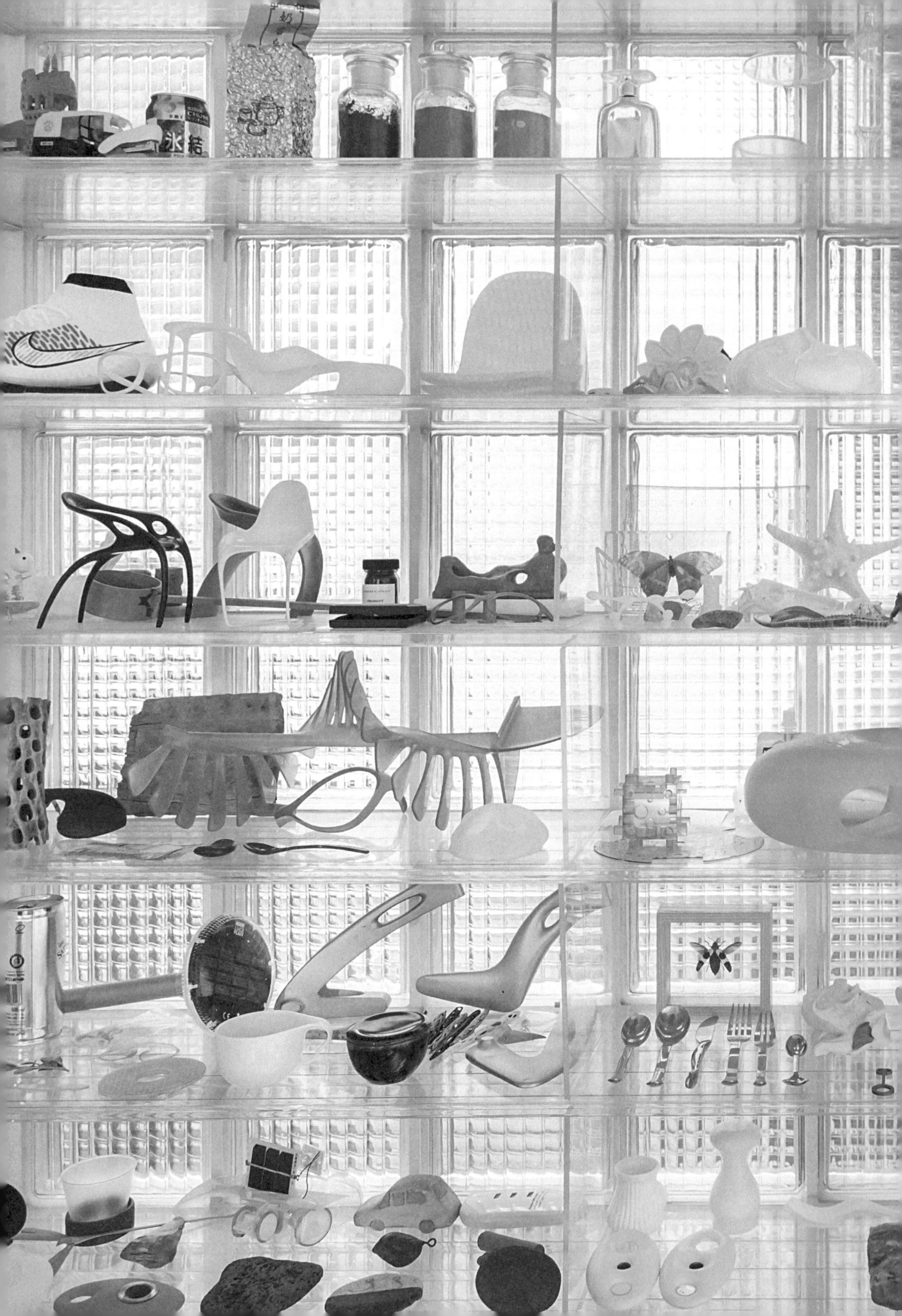
冰結